Series Editor's Foreword

Understanding the mechanisms of the reactions at transition metal sites is a key component in designing synthetic methods, developing industrial homogeneous catalysts, and investigating metalloenzymes. So this is an essential component in an undergraduate chemistry course.

Oxford Chemistry Primers are designed to give a concise introduction to all chemistry students by providing the material that would usually be covered in an 8–10 lecture course. As well as providing up-to-date information, this series will provide explanations and rationales that form the framework of an understanding of inorganic chemistry. Richard Henderson writes with the authority of a specialist and in a very approachable style. He covers the most important reaction classes from which one can build up more complex reaction sequences.

John Evans
Department of Chemistry, University of Southampton

Preface

Even if not everyone agrees that mechanistic studies are the jewels in the crown that is inorganic chemistry, it is undeniable that this topic represents an important course in the undergraduate syllabus. This book is aimed squarely at just such a course and covers those aspects of transition metal mechanistic chemistry that I consider essential. The book was written with the idea that it would act as a 'guide' and that the course lecturers will expand and elaborate on the topics covered and introduce the student to areas which I have not had space to address.

If this book manages to fire the imagination of even only a few students sufficiently that they read the more advanced texts on inorganic reaction mechanisms, or go on to do research in this area, it will have all been worthwhile.

My thanks go to Sundus and Matthew for spending so many fine spring weekends at home during the writing of this book; to Kay Oglieve for help in preparing the final script and, of course, to all my colleagues at Sussex for moulding the basic ideas and ideals on which this book relies. Last, but by no means least, I thank Martin Tobe for getting me addicted to mechanistic inorganic chemistry.

Brighton   R.A.H.
January 1993

The Mechanisms of Reactions at Transition Metal Sites

Richard A. Henderson

AFRC Institute of Plant Science Research, Nitrogen Fixation Laboratory,
University of Sussex

OXFORD NEW YORK TOKYO
OXFORD UNIVERSITY PRESS

Oxford University Press, Great Clarendon Street, Oxford OX2 6DP

Oxford   New York
Athens Auckland Bangkok Bogota Bombay
Buenos Aires Calcutta Cape Town Dar es Salaam
Delhi Florence Hong kong Istanbul Karachi
Kuala Lumpur Madras Madrid Melbourne
Mexico City Nairobi Paris Singapore
Taipei Tokyo Toronto Warsaw

and associated companies in
Berlin Ibadan

Oxford is a trade mark of Oxford University Press

Published in the United States
by Oxford University Press Inc., New York

A catalogue record for this book is available from the British Library

Library of Congress Cataloging in Publication Data
Henderson, Richard A.
The mechanism of reactions at transition metal sites / Richard A.
Henderson. — 1st ed.
(Oxford chemistry primers ; 10)
Includes index.
1. Transition metals. I. Title. II. Series.
QD172.T6H46   1993   546'.6—dc20   93–10835

ISBN 0–19–855746–9

Printed in Great Britain by
The Bath Press, Bath

Contents

1 Introduction to inorganic reaction mechanisms

1.1 A short history

In the period from 1920 to 1950, organic chemistry developed from a series of apparently unrelated facts to a discipline in which the application of a relatively few simple ideas concerning the electronic theory of reactivity resulted in an overall unified picture. This remarkably rapid growth was aided by several simplifying factors: the reaction centre (carbon) was the same in all studies; carbon has a single, stable oxidation state in these systems, a wide variety of compounds had already been prepared and structurally characterized, products were, in general, kinetically controlled and thus mechanistic information could be gained from product analysis experiments; and finally, most of the reactions were relatively slow and could be studied by conventional sampling methods.

In stark contrast, it was not until the 1950s that chemists started to look systematically at the reaction mechanisms of inorganic systems. The problem facing these early workers was immense: which of the hundred or so elements did they investigate? The problems did not stop there, since most of the elements can assume several different oxidation states, and each of these oxidation states can be associated with a wide range of coordination numbers, and each coordination number may be associated with more than one geometry! Nowadays, we would be spoilt for choice, but back in the 1950s the decision was driven by practicalities. First, in order to probe the mechanism meaningfully a wide variety of structurally characterized compounds were an important prerequisite. Secondly, the reactions had to be sufficiently slow to be studied by conventional sampling techniques, since this was before flow techniques for studying rapid reactions had been developed. Hence we see in the early days that, for instance, substitution mechanisms at octahedral sites were dominated by studies on the cobalt(III) complexes developed by Alfred Werner in the nineteenth century, and substitution mechanisms at square-planar sites were preoccupied with platinum(II) complexes, prepared primarily by the Russian researchers.

Now, of course, our knowledge of inorganic reaction mechanisms spreads across all areas of the periodic table, covering a wide range of reaction types, geometries and coligands (from classical coordination compounds through to organometallics and bioinorganic relevant studies). Inorganic reaction mechanisms is an active area of research and, having played their part in the early development of the subject, studies on the substitution reactions of cobalt(III) and platinum(II) complexes are now less common. Nevertheless it is these early studies which have established the patterns of reactivity around which the more contemporary studies are judged.

1.2   What is a mechanism and how is it determined?

The aim of this book is to show that the principles of mechanistic chemistry present a unifying description of the multitude of reactions that constitutes transition metal chemistry. However, before we rush headlong into describing the mechanisms of a wide variety of inorganic reactions it is essential that there is a clear understanding of what is meant by the word mechanism. Basolo and Pearson in their classic textbook on inorganic reaction mechanisms, last published in the 1960s, defined the word as, 'All the individual collisional and other elementary processes involving molecules that take place simultaneously or consecutively in producing the overall reaction'. Thus the mechanism of a reaction describes all the pathways by which a reactant is converted into products. This definition hides the complexity of the problem. If a reaction occurs by just one pathway, we have to define the way in which the reactant's ground state becomes activated, the structure of the derived intermediate and the manner in which this intermediate then collapses into product. Also, many reactions occur by several pathways, all of which require a description to the same degree of detail. Thus, mechanistic chemistry is necessarily preoccupied with the nature of the intermediates. Occasionally, these intermediates may attain sufficiently high concentrations to be detected spectroscopically and even more rarely, be sufficiently long-lived to be isolable. Under these circumstances the whole gamut of spectroscopic techniques can be used to probe the species. More frequently though, the intermediate is so short-lived that it cannot be detected directly; now the mechanist must use less direct methods to elucidate the nature of the intermediate. Clearly, if we have to employ these indirect methods then a large amount of circumstantial evidence must be accrued before the characteristics of the intermediate can be convincingly defined. During this process of defining a mechanism, two questions are paramount in the experimenter's mind. First, is the proposed mechanism consistent with all the known facts about the system under study? If it is not, then no amount of hand waving will change the fact that the mechanism is wrong. Secondly, once the experimenter is convinced of the mechanism, is there anything that the mechanism predicts which has not been investigated, such as the stereochemical consequences of the reaction?

Designing the system for a mechanistic study

It is clear that if we are to establish the detailed mechanism of any reaction unambiguously, then the system we choose to study must fulfil certain broad prerequisites, as shown in Fig. 1.1.

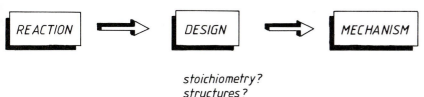

stoichiometry?
structures?
side reactions?

Fig. 1.1

Undoubtedly the most important prerequisite is that the structure of the reactants and the products must be known: determining the kinetics of a reaction in which the product (or even the reactant) is unknown is of no use in mechanistic chemistry. Obviously, a kinetic study cannot define the structure of the reactant. We should also make the system as simple as possible if we are to focus our attention on one aspect of the reaction system. For instance, consider the simple, generalized substitution reaction in Fig. 1.2.

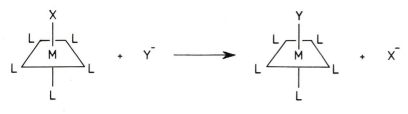

Fig. 1.2

If we wish to understand the factors which influence the simple substitution of X by Y^- we do not want secondary reactions involving the loss of the other ligands L to interfere. For this reason it is important to ensure that the remainder of the coordination sphere is essentially inert and merely acts as a spectator to the reaction under investigation. This may require a degree of fine-tuning of the system, involving synthetic chemistry. Finally we need the system to be amenable to perturbation. We need to vary the nature of the spectator ligands or the other reactants in a systematic manner to see how their various electronic and steric influences are reflected in the reactivity of the system.

Having designed the system which fulfils these prerequisites, we are now able to move on to the detailed mechanistic study, confident that we are targeting that aspect of the system of interest, uncomplicated by irrelevant side reactions. However, it is important to see Fig. 1.1 in its true light. Just because 'mechanism' occupies the far right-hand box in this figure does not mean that mechanism is the essence of chemistry. To be of true value the results of the mechanistic study should feed back pertinent information to chemistry in general, perhaps indicating new synthetic strategies or at least rationalize the reactivity of the system under investigation. Indeed the results may well have application beyond the realm of fundamental chemistry, and in particular over the last decade, the relevance of mechanistic inorganic studies to problems in bioinorganic chemistry have been much to the fore.

Currently it is fashionable to discuss studies on biological systems in books on inorganic reaction mechanisms. This will not be attempted in this short text. What does follow in the remainder of this book is a description of the fundamental reactions in transition metal chemistry. These fundamental reactions are the building blocks for understanding, at the atomic level, the multitude of reactions which make up chemical and biochemical reactions. The application of these principles to the understanding of the action of a particular metalloprotein is best left to specialist discussion of the pertinent protein.

1.3 Probing the mechanism

There are a variety of strategies which the mechanist can use to probe the mechanism of a reaction. In this section, only those that find the most general applicability will be outlined briefly to give the reader an idea of the approaches taken in the studies that make up the remainder of this book.

The rate law

The most important tool for the mechanist is the determination of the experimental rate law associated with the reaction. We are not going to devote space to the discussion of phenomenological kinetics or the techniques used to measure the kinetics of reactions as these are more than adequately covered in other textbooks in the reference section. Suffice it to say that the vast majority of inorganic kinetics are established employing the simplest approach: the method of pseudo first order conditions. In this analysis, the rate constant for the reaction is determined under conditions where all reagents are maintained in a large excess, compared to concentration of one of the reactants. By a series of such experiments, in which the concentrations of the species in excess are changed systematically, the complete rate law can be defined.

The interpretation of a rate law is reasonably straightforward. For the generalized reaction shown in Fig. 1.3 a variety of rate laws may be observed.

$$A \; + \; B \; \longrightarrow \; C$$

$$\frac{d[C]}{dt} = k[A][B] \quad \underline{or} \quad k'[A] \quad \underline{or} \quad k''[A][B]^2$$

Fig. 1.3

The rate law cannot be predicted by an inspection of the stoichiometry of the reaction, but only determined by experiment. The reaction may be an overall second order process: first order in the concentration of each reagent; an overall first order reaction: first order in the concentration of either A or B; or an overall third order process: for instance first order in A and second order in B.

The rate law gives the following information.
1. From the kinetic order: the number and types of each molecule participating in the formation of the transition state of the rate-limiting step.
2. From the rate constant: the rapidity of the reaction, allowing the comparison with analogous reactions.

In interpreting these parameters, the experimenter must arrive at a mechanism that is consistent not only with the rate law, but also with intuitive chemical sense. In general, the mechanism will consist of a series of elementary reactions which describe not only the interactions of the reactants prior to the rate-limiting step, but also those occurring after this slow step, which give rise to the products. The elementary reactions consist of unimolecular or bimolecular processes, since termolecular reactions are statistically unreasonable.

It is important to be clear that the kinetic order of the reaction is not the molecularity of the reaction. As an example consider the system in Fig. 1.3. The third order kinetics clearly do not mean that the reaction is termolecular, rather that in the series of elementary steps leading to the transition state, two molecules of B sequentially interact with one molecule of A. It is important to remember that the kinetic order is determined experimentally, whilst the molecularity of a reaction is a mechanistic interpretation.

Unfortunately it is a fact of life that the simpler the rate law observed for a reaction, the less certain you can be about the mechanism to which it relates. Consequently many kinetic studies are aimed at trying to perturb the simpler kinetics in order to limit the number of mechanisms with which they are consistent.

As an example of the sort of problems one encounters in interpreting a rate law, consider the studies on the substitution reactions of cobalt(III) complexes (Fig. 1.4).

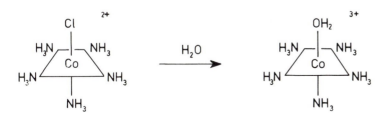

Fig. 1.4

The rate of this reaction shows only a first order dependence on the concentration of the cobalt complex and by analogy with the kinetics of the S_N1 reaction at a saturated carbon atom, it could be assumed that this reaction was also dissociative. This in essence is true, but these simple kinetics do not suffice in defining that mechanism. After all, what else would one expect for the kinetics of this reaction? The nucleophile (water) is also the solvent present in such a large excess that any participation by it in the activated complex could not be monitored. In this case the experimenter must take recourse to other parameters which can discriminate between unimolecular dissociation of the cobalt–chlorine bond, and bimolecular attack of water at the cobalt centre.

Another reaction showing simple kinetics is the substitution of the dinitrogen ligand in *trans*-[Mo(N$_2$)$_2$(Ph$_2$PCH$_2$CH$_2$PPh$_2$)$_2$] shown in Fig. 1.5. This reaction exhibits a simple first order dependence on the concentration of the molybdenum complex provided the nucleophile is present in a large excess. The observed rate constant under these circumstances is the same, irrespective of the nature of the nucleophile (RCN, RSH or RI), consistent with rate-limiting dissociation of dinitrogen k_1.

However, at relatively low concentrations of the nucleophile the true rate law is sometimes revealed and is significantly more complicated than that described so far, as shown by the expression in Fig. 1.5. This generalized rate law is derived by treating the five-coordinate coordinatively unsaturated intermediate as a steady state species. At sufficiently low concentrations of nucleophile the rate exhibits a first order dependence on the concentration of that nucleophile, but at high concentrations of nucleophile $k_2[RCN] \gg k_{-1}$ and the rate of the reaction becomes independent of both the nature and concentration of the nucleophile.

In this latter example, the complexity of the rate law permits a detailed mechanism to be established with a reasonable degree of confidence as to its correctness. The important feature about the derivation of the

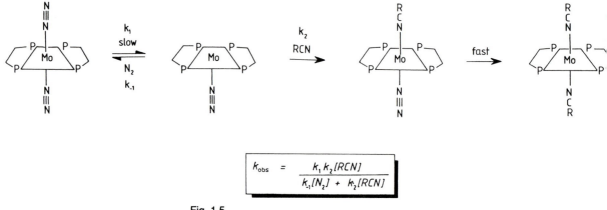

$$k_{obs} = \frac{k_1 k_2 [RCN]}{k_{-1}[N_2] + k_2[RCN]}$$

Fig. 1.5

mechanism from the form of the rate law is that only secondary notice is taken of the magnitude of the derived rate constants. Although accurately establishing the numerical value of the rate constant is of importance, we must always bear in mind that this is not an end in itself, and that just accurately measuring the rate constant for a simple, unambiguous process is often of very limited mechanistic value.

Extra-kinetic parameters

In addition to the determination of the rate law at a single temperature, it is sometimes of mechanistic value to study the temperature dependence of a reaction. Over a relatively narrow temperature range, the rate constant changes in accordance with the Eyring equation shown in Fig. 1.6.

Derivation of the corresponding values of $\Delta H^{\ddagger}$ and $\Delta S^{\ddagger}$, the enthalpy and entropy of activation, for elementary steps in a reaction mechanism give information about whether differences in rate constants for apparently analogous reactions are attributable to bond strength differences (reflected in $\Delta H^{\ddagger}$) or, for instance, specific solvation effects (reflected in $\Delta S^{\ddagger}$). In particular, the determination of whether $\Delta S^{\ddagger}$ is positive or negative has been used extensively in the study of substitution reactions to discriminate between dissociative and associative mechanisms. However, as we have already seen, many of the reaction mechanisms are multistep processes with rate laws which are only limiting forms of more complex equations. Consequently, the observed rate constant may in reality correspond to a complicated quotient of elementary rate constants. Measuring the temperature dependence in such a situation is often of limited value since all it tells you is the overall effect of temperature on the combined rate constants.

In some cases, the influence of temperature on the multistep nature of the mechanism can be used to advantage. For instance, in the reaction shown in Fig. 1.7, varying the temperature over a relatively modest range gives a distinctly curved Eyring plot.

$$k = \frac{kT}{h}.exp\frac{\Delta H^{\ddagger}}{RT}.exp\frac{\Delta S^{\ddagger}}{RT}$$

$k =$ *Boltzmann's constant*
$h =$ *Planck's constant*

Fig. 1.6

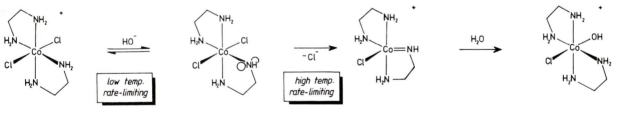

Fig. 1.7

This curved behaviour is consistent with a change in the rate-limiting step from deprotonation to dissociation of chloride. Clearly this sort of behaviour will only be evident if the values of $\Delta H^{\ddagger}$ for the two steps are significantly different.

A word of caution about Eyring plots is pertinent at this stage. There is no point in measuring these temperature dependencies over extended temperature ranges (a range of 20°-30°C is normal), since there is no good reason why the Eyring equation should be valid over a protracted range, and hence any observed curvature may have no mechanistic significance.

A relatively new parameter which is finding application in a wide variety of mechanistic studies is the volume of activation, $\Delta V^{\ddagger}$, determined by monitoring the effect of pressure on the rate of the reaction. Caution should be exercised when interpreting $\Delta V^{\ddagger}$ since it is made up of two components. The two components are:

(1) the intrinsic volume change of going from reactants to the transition state, $\Delta V^{\ddagger}_{int}$;
(2) the volume change associated with solvation effects, $\Delta V^{\ddagger}_{sol}$.

The term of most interest to most mechanists is $\Delta V^{\ddagger}_{int}$, since it relates to changes in bond lengths and geometry upon reaching the transition state. It is not always easy to estimate the relative contributions of $\Delta V^{\ddagger}_{int}$ and $\Delta V^{\ddagger}_{sol}$ to $\Delta V^{\ddagger}$.

Electronic and steric effects

These two effects are treated together here since it is often difficult to separate one from the other.

One example where steric factors do undoubtedly operate is in the redox reactions of the complex shown in Fig. 1.8.

In the oxidation of these nickel(I) complexes by substitutionally inert cobalt(III) complexes, the rate varies little with the nature of the quadridentale ligand $k_{tmc}/k_{dmc} \simeq 4$. However, in the reactions with other oxidants such as alkyl halides or peroxides there is a large difference in the rate of reaction with the two complexes, $k_{tmc}/k_{dmc} = 104$–105. The greater sensitivity of the latter reactions to the substituents is due to these reactants having to bind to the nickel prior to oxidizing it, whereas the inorganic oxidants do not.

In general, it is easier to establish the electronic influence of substituents on the reaction rate, free from the complications of steric effects. Here, as in organic reactions, extensive use is made of Hammett functions with the substituent far removed from the reaction centre. Often it is the departures from the expected Hammett correlation that gives the most interesting mechanistic insight. For example in the substitution reaction shown in Fig. 1.9, there is a good linear correlation between the carbonyl stretching frequency, ν_{co} and the Hammett σ (for the substituent R on the cyclopentadienyl-ring) in both the reactant and the products. The correlation with the rate of the reaction though is not so good.

In particular there is a marked deviation from the Hammett σ value for the substituents NMe_2 and Cl. Since there is a good correlation with the

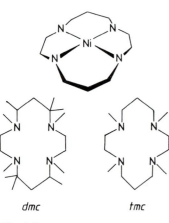

dmc   *tmc*

Fig. 1.8

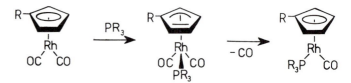

Fig. 1.9

carbonyl stretching frequencies (which reflects the electron-density at the rhodium atom through back-bonding), the variation observed in the rate is reflecting an effect on the transition state of the reaction. It has been proposed that the lone pairs of electrons on the substituents can become delocalized over the cyclopentadienyl ring which is only coordinated in an η^3 mode in the transition state of an associative mechanism.

1.4 A classification of inorganic reactions

If we consider a simple metal complex as shown in Fig. 1.10, consisting of a metal surrounded by a number of ligands in a defined geometry (in this case an octahedron), there are three general classes of reactions which can occur.

(1) the addition or removal of a ligand *substitution*;
(2) change in the oxidation state of the metal *electron transfer*;
(3) attack at a ligand *activation of ligands*.

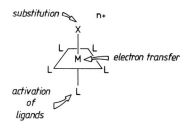

Fig. 1.10

These three categories make up the main sections of this book, and are the principal reactions associated with transition metal chemistry. Combinations of these processes lead to other well-known reactions typical of these sites. Thus a combination of categories (1) and (2) make up the area of *redox-catalysed substitution* (Chapter 5) and *oxidative-addition* (Chapter 7). A combination of categories (1) and (3) make up *insertion reactions* (Chapter 9), and finally a combination of categories (2) and (3) is associated with *redox-catalysed insertion reactions* (Chapter 9).

The wide range of metal complexes that make up any discussion of inorganic reaction mechanisms means that a variety of reaction rates will be observed, ranging from the so-called 'instantaneous', through to those that take several days to complete. The general terminology to describe these widely different reactivities is that sites undergoing very rapid reactions are termed *labile*, while the sites that react sluggishly are termed *kinetically robust* or *inert*. The best working discrimination between these two types of complexes is that a labile reaction is one that is complete within the time of mixing the two reagents.

PART ONE: SUBSTITUTION REACTIONS

2 Substitution mechanisms at transition metal sites

The detailed understanding of substitution processes relates to several areas of chemistry, for instance, the binding of substrates to the active site in both metalloproteins and homogeneous catalysts. It is also an important feature in many electron transfer reactions.

In deciding how to present the substitution mechanisms of transition metal centres, there is the problem of assessing which is the most logical approach. Here, as in other texts, the substitution mechanisms will be discussed in terms of the coordination number and geometry of the metal site. Although this is probably the best strategy it has certain limitations which are briefly discussed below. Knowing these limitations the parallels and analogies between mechanisms at very different reaction sites, that are distributed throughout the book can be more fully appreciated.

The discussion of substitution mechanisms in terms of coordination number has as its main proposition that each coordination type has a particular group of mechanisms which, on the whole, are typical of it, but are different to those observed with other coordination numbers. It is now becoming increasingly clear that this is not the case. Both associative and dissociative mechanisms are now well established in four-, five-, six-, and seven-coordinate complexes. Also, the factors which dictate the relative contributions of these two extreme mechanisms in all types of complexes are the same. There is, however, a more profound problem even within the discussion of, for instance, substitution mechanisms of six-coordinate complexes. Even the most casual of readers would appreciate that there are distinct differences between a complex such as $[CoCl(NH_3)_5]^{2+}$ and $[Mo(CO)_6]$. The former contains cobalt (a first row transition metal in group IX) in the +3 oxidation state, is a 'hard' metal centre with the corresponding set of 'hard' ligands, whereas the latter contains molybdenum (a second row transition metal in group VI) in the zero oxidation state, is a much 'softer' metal centre with its corresponding carbonyl ligands. Yet we want to discuss the mechanisms of these two species at the same time, even though they are not soluble in the same solvent. Despite all this though, they do have many mechanistic features in common.

Back in the early days of inorganic reaction mechanism studies it was contended that the reaction mechanisms of each system were peculiar to itself and that there was no reason why there should be common features, even for the same metal with various coligands. From the descriptions that follow in the remainder of this book, it will become clear that this proposition is not true. There is no point in establishing the principles of a reaction mechanism for one type of system only to find that an entirely different set of principles operates in an analogous system. Mechanistic studies must aim at establishing generalities to be of value to chemistry in general.

2.1 General considerations

In considering any substitution mechanism in very general terms it is clear that there are three classes of mechanism, as shown in Fig. 2.1.

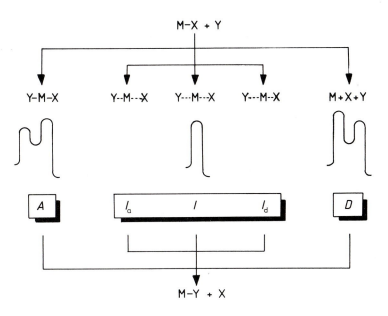

Fig. 2.1

Following the nomenclature devised by Langford and Gray, these mechanisms can be categorized as follows. On the far left hand side is the associative mechanism (A) in which the entering group binds to the reaction centre before any bond weakening of the leaving group has occurred. In this mechanism the reaction profile consists of a single intermediate (containing the entering and leaving group bonded to the reaction centre), and two transition states leading to the formation and decomposition of this intermediate. On the far right-hand side of Fig. 2.1 is the dissociative mechanism (D) in which the leaving group has left the reaction centre prior to any interaction with the entering group. This mechanism has a reaction profile consisting of a single intermediate and two transition states. Now, the intermediate has a lower coordination number than either the reactant or the product. The difference between the profiles for the A and D mechanisms is the relative barrier heights on either side of the intermediates. In the A mechanism the barrier to the decomposition of the intermediate is the larger of the two, whereas in the D mechanism, the formation of the intermediate has the larger barrier height.

In the centre we see the third possibility, namely that the loss of the leaving group and the approach of the entering group are synchronous. In this mechanism, as the entering group approaches the reaction centre, the leaving group responds in a concerted manner and progressively leaves the site, until the point when the entering group becomes tightly bound and the leaving group is entirely dissociated. This so-called interchange mechanism (I) necessarily has a reaction profile containing no intermediate, only a single transition state.

Of course these three designations represent only the two extremes and the dead centre of all the possible substitution types that can operate. In reality there is a spectrum of mechanism types which covers all the subtle variations between the I and A mechanisms on the one hand, and the I and D mechanisms on the other. In order to account for these grey areas, and to give them some formal designation, they have been termed I_a and I_d, respectively. These mechanisms are best considered as perturbations of the I mechanism.

For the I_a mechanism, the leaving group does not respond sufficiently rapidly to the approach of the entering group and thus retains a tighter hold on the reaction centre than it does in the I case, right up to the time when the entering group becomes completely bound to the site. On the other hand, in the I_d mechanism the leaving group is very sensitive to the approach of the other group and readily relinquishes its hold on the reaction centre, even before the entering group has become tightly bound. It can be seen that these subtle variations to the identity of the reaction mechanism are associated with the relative heights of the barriers, and hence the relative energies of the transition states.

Operationally it is relatively easy to establish that a mechanism proceeds through a D or A mechanism, when the intermediate attains sufficiently high concentration to be detected. However, it is not easy to be so confident as to what the exact intimate mechanism is, in cases where the intermediate is less willing to reveal itself. In the remaining discussions it must be remembered that all pathways can operate in all systems, at least in principle, and which pathway dominates depends on the relative barrier heights associated with the various pathways.

3 Substitution at four-coordinate sites

3.1 Substitution at a tetrahedral metal site

There are few transition metal sites associated with a coordination number less than four and hence we will start our discussion of substitution mechanisms by looking at tetrahedral metal sites. This has the advantage that the reader is presumably familiar with an analogous discussion at saturated carbon atoms, but more importantly, there are some general principles which, once established in this section, will be useful for the remaining chapters.

The reactions of the coordinatively saturated, tetrahedral complexes (those with a formal 18 electron count) such as $[Ni(CO)_2(PR_3)_2]$, $[Ni(CO)_4]$ or $[Ni\{P(OEt)_3\}_4]$ all exhibit very simple kinetics: a first order dependence on the concentration of the metal complex, and an independence of the concentration and nature of the reacting nucleophile.

These simple kinetics have been interpreted in terms of a dissociative mechanism (D) as shown in Fig. 3.1, and this conclusion is substantiated further by extra kinetic parameters. Thus in the reaction of $[Ni(CO)_4]$ with PEt_3 the value of $\Delta V^{\ddagger} = +8$ cm^3 mol^{-1} has been determined.

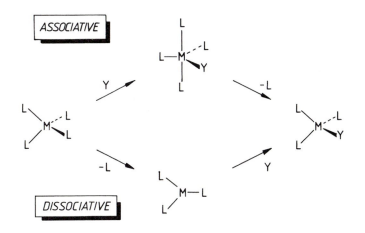

Fig. 3.1

In contrast, the phosphine exchange of $[CoBr_2(PPh_3)_2]$ (15 electron species) and halide exchange of $[FeBr_4]^-$ (13 electron species) both show second order kinetics: first order in the concentrations of the complex and the nucleophile. The associative (A) nature of these reactions is further amplified by the large negative values of $\Delta S^{\ddagger}$ and, for the cobalt system, the value of $\Delta V^{\ddagger} = -12.1 \text{ cm}^3 \text{ mol}^{-1}$.

One might be tempted to predict that the 18 electron species, $[Co(CO)_3(NO)]$ and $[Fe(CO)_2(NO)_2]$, would react by a dissociative pathway. However, the rates of reaction of these nitrosyl systems are sensitive to both the concentration and nature of the nucleophile, characteristic of associative (A) mechanisms. This behaviour is due to the presence of the nitrosyl ligand. This ligand is capable of binding to a metal centre either by acting as a three electron donor (in which case the M–NO fragment is linear) or as a one electron donor (in which case the M–NO fragment is bent). Hence, at least in principle, nitrosyl complexes can readily become coordinatively unsaturated by simple intramolecular change in the nature of the metal nitrosyl interaction. The generated 16 electron species is then susceptible to attack by nucleophiles as shown in Fig. 3.2.

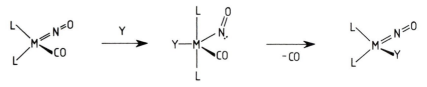

Fig. 3.2

This is not a type of activation which is restricted to tetrahedral substitution, and we shall see this behaviour associated with higher coordination numbers and other flexible ligands.

A simple conclusion then is that the lowest energy pathways for substitution at coordinatively saturated, tetrahedral complexes are those involving 16 electron intermediates. But is this always true? Most mechanistic studies are performed relatively close to 25°C. What happens to the relative energies of the associative and dissociative pathways as we move to more extreme temperatures? The answer to this problem is addressed by considering the substitution mechanisms of $[Ni(CO)_3(N_2)]$ in liquid krypton (T = -160°C), as shown in Fig. 3.3.

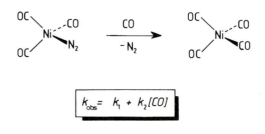

$$k_{obs} = k_1 + k_2[CO]$$

Fig. 3.3

Although this complex is isoelectronic with $[Ni(CO)_4]$ which, as we have seen, reacts by a purely dissociative pathway at 25°C, the kinetics of the reactions of $[Ni(CO)_3(N_2)]$ in liquid krypton demonstrate that both associative and dissociative pathways operate. When a reaction can proceed by two pathways, which one is the dominant pathway depends critically on the temperature, since it is highly unlikely that each pathway responds to changes in the temperature by identical amounts. Inspection of the free energies for the barriers for the reactions of $[Ni(CO)_3(N_2)]$ shows that the ratio, k_1/k_2 changes by about five orders of magnitude over the temperature range 25° to −160°C. Thus, over this temperature range the relative energetics of the associative and dissociative pathways changes. If we were able to study the substitution reactions of $[Ni(CO)_3(N_2)]$ at 25°C we would find that the exceedingly rapid reactions operated exclusively by a dissociative mechanism.

3.2 Substitution at a square-planar site

Rate law and mechanism

The substitution reaction at a square-planar site is one of the most extensively studied areas in inorganic reaction mechanisms, with most of the information established at a platinum(II) centre. However, other d^8 sites such as Ni(II), Pd(II), Au(III), Rh(I), and Ir(I) have been investigated, though less extensively.

If one considers the basic substitution reaction such as that shown in Fig. 3.4, then usually the rate law observed exhibits a mixed order dependence.

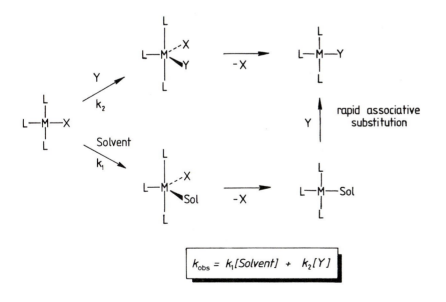

$$k_{obs} = k_1[Solvent] + k_2[Y]$$

Fig. 3.4

By comparison with the discussion on the substitution mechanism at tetrahedral sites, one might expect a parallel dissociative (k_1) and associative (k_2) pathways. However, when the reaction is performed in a solvent which is capable of acting as a nucleophile, there remains the possibility that both pathways are associative with the k_2 route representing direct attack by the nucleophile at the platinum centre, and the k_1 route involving rate-limiting attack by the solvent. The solvent ligand is only relatively weakly held and is subsequently rapidly replaced by the nucleophile in a further associative pathway.

There is a great deal of circumstantial evidence that the true mechanism is that shown in Fig. 3.4 with both pathways being associative. For instance the values $\Delta S^{\ddagger} = -50$ to -180 J K^{-1}mol^{-1} and $\Delta V^{\ddagger} = -6.4$ to -10 cm^3 mol^{-1} for the reactions of [Pt(dien)Br]$^+$ (dien = H$_2$NCH$_2$CH$_2$NHCH$_2$CH$_2$NH$_2$) strongly implicate an associative mechanism, probably I$_a$. Similarly, the repeated appearance of a two term rate law in the reactions of square-planar complexes indicates that both pathways are similarly sensitive to the same factors. In addition, performing the reactions in a poorly coordinating solvent leads to a rate law exhibiting only the nucleophile (k_2) pathway. Increasing the steric congestion at the reaction centre by introducing bulky coligands decreases the rate of both the k_1 and k_2 pathways by similar amounts. This is best illustrated in the extensive studies on the reactions of the series of complexes, [PdCl(R$_5$dien)]$^+$ (R$_5$dien = R$_2$NCH$_2$CH$_2$NRCH$_2$CH$_2$NR$_2$). These complexes react with I$^-$ showing the normal two term rate law, k$_{obs}$ = k_1[solvent] + k_2[I$^-$], but as the steric bulk around the metal is increased by varying the tridentate ligand (from H$_5$dien to Et$_5$dien), the rate constants associated with both pathways progressively decreases, such that the Et$_5$dien complex reacts about 1×10^5 times slower than the H$_5$dien species, and the k_2 pathway becomes increasingly difficult to observe. It is often difficult to distinguish between true steric effects and electronic influences of the increasing alkyl substituents, or even a change in the mechanism through the series. However, measurement of the volume of activation for these systems shows that $\Delta V^{\ddagger} = -12.2$ cm^3 mol^{-1}, and hence the reaction remains associative throughout the series.

The solvento-intermediate

A key feature of the two term rate law is the involvement of a labile solvento-species as an intermediate in the k_1 pathway. Several studies on isolated solvento-species, such as [Pt(dien)OH$_2$]$^{2+}$, demonstrate that these species react rapidly with nucleophiles. This criterion alone is not sufficient to prove that these species are intermediates in the mechanism. However, the intermediacy of the solvento-species has been demonstrated by trapping the aquo-species with hydroxide ion, as shown in Fig. 3.5.

The reaction obeys the normal two term rate law with the assumed aquo-intermediate involved in the k_1 pathway. The trick employed in the trapping of this intermediate is the deprotonation of the aquo-ligand by hydroxide ion to form the hydroxy-species. This deprotonation involves no

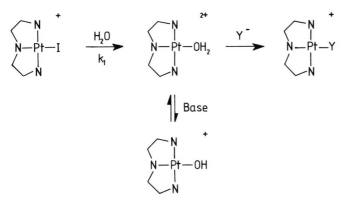

Fig. 3.5

metal–oxygen bond cleavage and is very much more rapid than the substitution of the coordinated water by the nucleophile, Y^-. In addition the derived hydroxy-species is essentially inert to substitution. By investigating the kinetics of the reaction in the presence of various concentrations of hydroxide ion, it is observed that the substitution of the iodo-group by the nucleophile is inhibited by hydroxide ion.

Stereochemistry of the substitution

The substitution reactions of the vast majority of square-planar complexes proceed with retention of stereochemistry: *trans* complexes give *trans* products and *cis* complexes give *cis* products exclusively. This stereochemical integrity is most readily rationalized in terms of a trigonal bipyramidal intermediate, as shown in Fig. 3.6.

Fig. 3.6

In this description the nucleophile approaches the metal centre along a line perpendicular to the four ligands, that is along a line of minimized steric interference. As the nucleophile approaches the metal, the leaving group starts to weaken its hold on the platinum and distorts the regular geometry by bending away from the plane containing the *trans* ligand and the two *cis* ligands. In this way, the trigonal plane of the five-coordinate intermediate always contains the leaving group, the attacking nucleophile and the group originally *trans* to the leaving group. The two groups that were originally *cis* to the leaving group occupy the axial positions. Subsequent loss of the leaving group allows the, newly coordinated nucleophile to relax smoothly into the plane containing the other three ligands.

There are few exceptions to the stereochemical observations outlined above. When exceptions are observed the origin of the effect can be traced to three possible causes.

1. There is a relatively low barrier to intramolecular rearrangement between the square-planar and the tetrahedral form. As can be seen in Fig. 3.7, in the tetrahedral form the stereochemical relationship between the various ligands in the square planar forms is lost. This type of intramolecular rearrangement can clearly occur in either the reactant or the product.

2. Stereochemical integrity can be lost in the intermediate, as shown in Fig. 3.7. If the trigonal bipyramidal intermediate is sufficiently long-lived, and the propensity of the axial ligands to enter the trigonal plane is high, then the intermediate may undergo an intramolecular pseudorotation process which will exchange the axial ligands with two ligands in the equatorial plane. Several of these exchanges result in the complete loss of stereochemical integrity of the trigonal bipyramid. The only restriction in the ensuing product-forming dissociation is that the leaving group is lost from the equatorial plane, but what else this plane contains will depend on the pseudorotation process.

3. A departure from the usual stereochemical pattern is observed if the substitution becomes dissociative, which will be discussed later.

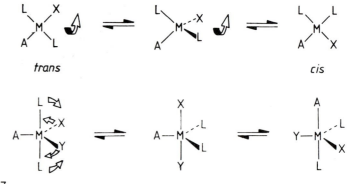

Fig. 3.7

Influence of the entering group

In any substitution reaction which operates by an associative mechanism it is possible, at least in principle, to set up a scale of the relative abilities of each nucleophile to accomplish substitution at the reaction centre. Of course this will not be a universal scale, the relative effects of each nucleophile is dependent upon the reaction centre being attacked. For instance, the well-known S_N2 reaction at a saturated carbon will have a different scale of nucleophilicity than that for the platinum centre of most interest here. The most obvious difference between the two is that the saturated carbon centre is a 'hard' centre, being a first row p-block element, whereas the platinum(II) centre is much more 'soft', being a third row d-block element. The relative rates of nucleophilic attack will reflect the differing affinities of the nucleophile for these two sites.

In line with other scales of nucleophilicity, a scale has been established for platinum(II) sites with the nucleophilicity, n_{Pt}, given by the equation in Fig. 3.8, for the reactions of *trans*-[PtCl$_2$py$_2$].

The relative nucleophilicities reflect the 'soft' centre of platinum. Thus the heavier halides react more rapidly with this centre than their lighter analogues, as do sulphur compounds than their oxygen analogues, or arsenic compounds than their nitrogen analogues.

For general applicability it is essential that this nucleophilicity scale be related to the rates of other platinum reactants. This has been accomplished with some success using the equation shown in Fig. 3.8. Here the parameter $n_{Pt}{}^0$ (n_{Pt} corrected to zero ionic strength) is used. This empirical equation is associated with the two parameters S which is the gradient of the best straight line through the data, and C the intercept on the log k_Y axis.

The parameter S is called the nucleophilic discrimination factor and reflects the response of the reaction rate to the nucleophilicity of the attacking group and the centre being attacked. The parameter C is called the intrinsic reactivity and is much more difficult to assign a physical meaning. It can be shown that this parameter is the first order rate constant for the solvolytic pathway, were that reaction to be carried out in methanol at 30°C at zero ionic strength. However this does not help too much in understanding the physical origin of this parameter.

Although the nucleophilicity scale described above works adequately for neutral complexes, when we consider cationic complexes the same scale is not strictly obeyed. The problem is compounded by the fact that new scales are required for monocations, dications, monoanions, etc. This shows the importance of charge in determining the nucleophilicity in these systems, but does tend to negate the utility of such scales.

$$n_{Pt} = \log(k_Y/k_{MeOH})$$

$$\log k_Y = Sn_{Pt}^0 + C$$

Fig. 3.8

Influence of the leaving group

There is very little to be said about the influence of the leaving group on the rate of the reaction. Often the ability of a ligand to be a good leaving group parallels the bond strengths of that residue. However, in any associative mechanism, the attacking nucleophile is present in the coordination sphere of the metal when the leaving group departs and this can clearly influence the rate of the dissociation.

Influence of the spectator ligands

We have already seen how the steric effects of the spectator ligands influence the reaction rates of square-planar complexes, we will now look at the electronic effects of these ligands. Stereochemically, there are two different types of ligands in a square-planar complex: the two that are *cis* to the leaving group and the single residue that is *trans* to the leaving group. It has been established that the nature of the *trans* ligand plays the dominant role in defining the product. The empirical observation is best explained by reference to the generalized reaction shown in Fig. 3.9.

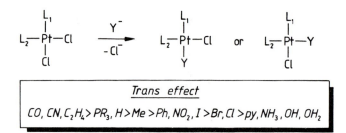

Fig. 3.9

There are two possible products from the reaction with the general nucleophile, Y^-, depending on whether the chloro-group *trans* to L_1 or L_2 is replaced. What is observed is that the proportion of product depends upon the nature of each L, and that the effect is dominated by the *trans* ligand (not the *cis*), and is independent of the nature of the nucleophile. The effect of the ligands on the rate of substitution of the group *trans* to itself is known as the *trans* effect.

The order of the common ligand's effect on the rate of substitution of the *trans* ligand is shown in Fig. 3.9. Simple inspection of this list reveals that the nature of the groups with a high *trans* effect is very disparate. Groups such as hydrides, alkyls and ammonia, which can only operate as simple σ-donors are spread throughout this list, as are groups such as carbon monoxide, tertiary phosphines and pyridine which can act as reasonable π-electron acceptors. In order to understand the various effects operating in the *trans* effect series at an atomic level we have to consider not only effects in the ground state of the reactant, but also effects that may operate in the transition state of the reaction. Thus an increase in rate of one reaction over another can arise from any factor making the activation barrier lower: either a destabilization of the ground state or a stabilization of the transition state. Both factors operate to differing degrees throughout the members of the *trans* effect series.

Ground state destabilization

This effect operates with the pure σ-donors such as hydride and alkyl, in which these groups are strongly polarizing and pull electron density towards themselves thus depriving the other ligands, and most particularly the group *trans* to itself, of electron density. This weakens the *trans* ligand's binding to the metal and hence increases its lability.

Transition state stabilization

With groups such as coordinated alkenes, carbon monoxide, tertiary phosphines, etc., relatively low-lying vacant orbitals are available which can be used to form multiple bonds by accepting electron density from the metal centre. This is of most utility in the transition state where the metal has a potential buildup of electron density while the entering nucleophile and the leaving group are both bound to the metal. By making use of these vacant orbitals on the *trans* ligand at least some of this excessive electron density can be dissipated away from the metal, as shown in Fig. 3.10.

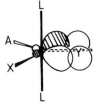

Fig. 3.10

It must be remembered that the *trans* ligand is the only group in the trigonal plane of the trigonal bipyramidal intermediate with the leaving group and the entering group, and so is uniquely positioned to accept the π-electron density.

3.3 Dissociative mechanisms at square-planar sites

It is not easy to demonstrate that the reaction of a square planar complex operates via a dissociative pathway, rather than the usual associative routes we have discussed above. Clearly the demonstration that the rate of the reaction is independent of the concentration and nature of the nucleophile is ambiguous in this type of system; it may correspond to a dominant pathway involving attack by the solvent on the complex (the k_1 pathway in Fig. 3.4).

Dissociative mechanisms at the square planar complexes, *cis*-$[PtR_2(OSMe_2)_2]$ (R = Me or Ph) and *cis*-$[PtMe_2(SMe_2)_2]$ in their reactions with a range of bidentate ligands have been demonstrated unambiguously. With these complexes the complicated rate law, shown in Fig. 3.11, shows that the substitution proceeds by both an associative pathway (k_2) and, most importantly, a dissociative pathway.

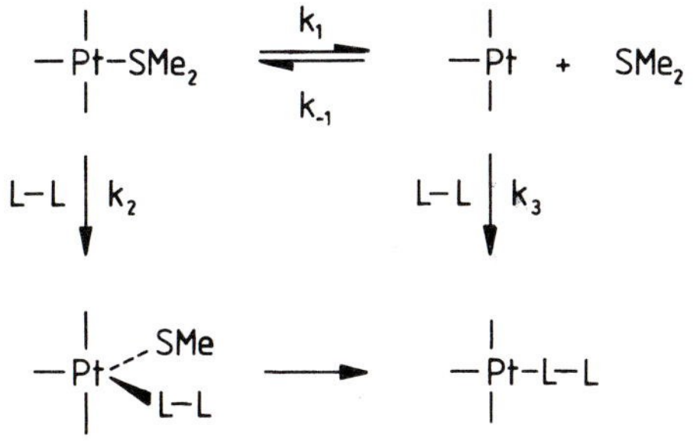

$$k_{obs} = \frac{k_1 k_3 [L-L]}{k_{-1}[Me_2S] + k_3[L-L]} + k_2[L-L]$$

Fig. 3.11

This first term in the rate law is consistent only with a dissociative route: an independence of the rate on the concentration and nature of the nucleophile in the high concentration limit, and the reaction rate inhibited by the addition of an excess of the leaving group. An analogous rate law for the substitution of coordinated acetonitrile in *trans*-$[RhCl(NCMe)(PPh_3)^2]$ by PPh_3 to give $[RhCl(PPh_3)_3]$ indicates that this species also reacts, in part, by a dissociative mechanism.

The three coordinate intermediate

The dissociation of a ligand from a square-planar complex, as shown in Fig. 3.11, gives rise to a three-coordinate, 14 electron, T-shaped intermediate, as shown in Fig. 3.12.

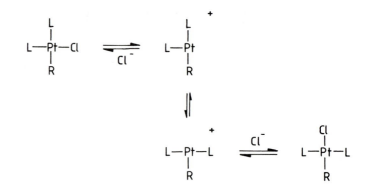

Fig. 3.12

Intramolecular rearrangements of the 'cis-like' T-shaped intermediate into the 'trans-like' T-shaped intermediate, and subsequent capture by a nucleophile can result in non-stereospecific reactions.

4   Substitution at sites with coordination number greater than four

Undoubtedly the most common coordination number in transition metal chemistry is six. Consequently a great deal is known about the substitution mechanisms at octahedral sites, but significantly less about these same processes at five- or seven-coordinate sites. We shall see in this chapter that dissociative mechanisms dominate the substitution reactions at six-coordinate species, but associative mechanisms can, in certain circumstances, also operate. There is no need, at least on a mechanistic basis, to differentiate between the classic Werner-type complexes, such as $[CoCl(NH_3)_5]^{2+}$, and the systems which fall in the realm of organometallic chemistry, such as $[Mo(CO)_6]$. However, it is true that the detailed understanding of substitution mechanisms is greatest for the six-coordinate, Werner-type cobalt(III) complexes, and it is only relatively recently that the factors influencing the substitution mechanisms at organometallic-type sites has received attention. The reason for this is largely historical, in that the Werner-type complexes were the first to be investigated back in the early 1950s, when organometallic chemistry was far from developed.

The factors which discriminate between associative and dissociative mechanisms in six-coordinate complexes should be the same as those that operate for any coordination number. In particular we will see that for organometallic systems, a few simple principles allow us to understand which mechanism will dominate for coordination numbers from four to eight.

The role of the solvent

We saw in the studies on square-planar complexes that the solvent is actively involved in the substitution mechanism. The solvent also plays an important role in the substitution mechanisms of octahedral complexes, but in a different way. If the generalized substitution reaction shown in Fig. 4.1 is considered at the crudest level of mechanistic sophistication, it is clear that the reaction can occur by direct substitution of coordinated X for Y or by the initial replacement of coordinated X for solvent followed by the replacement of the solvent molecule by Y.

This latter pathway is of most importance when the solvent has a reasonably high affinity for the metal site, and indeed dominates the substitution characteristics of the Werner-type complexes performed in

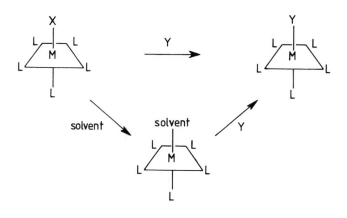

Fig. 4.1

water. In contrast, the substitution of the relatively low oxidation state complexes is often performed in solvents with relatively low coordinating power, and hence the reaction proceeds without the intermediacy of a solvento-species.

It is important to bear in mind that typically the kinetics of these substitution reactions would be studied at only millimolar concentrations of Y, whereas the solvent would be present in several molar concentration (for example, $[H_2O] = 55.5$ mol dm^{-3}, $[MeOH] = 31.3$ mol dm^{-3}) and thus is present in such a large excess that it will invariably win in the competition for the vacant site on the metal centre even if it is only a relatively weak nucleophile.

Clearly, to build up a detailed picture of substitution behaviour at six-coordinate sites we need to understand the details of all the steps shown in Fig. 4.1:

(1) the solvolysis reaction;
(2) the displacement of coordinated solvent;
(3) the direct replacement of one group for another without the intermediacy of a solvento-species.

4.1 The solvolysis reaction

In defining the details of the solvolysis reaction we are interested in a common type of entering group, the solvent. One problem in studying this type of reaction is that any kinetic analysis will not reveal the dependence of the reaction rate on the concentration of the nucleophile, the solvent.

The most comprehensive accumulation of data for this type of reaction comes from the studies of classical Werner-type cobalt(III) complexes. For the generalized substitution reaction shown in Fig. 4.2, the reaction is invariably associated with a positive $\Delta S^{\ddagger}$ and $\Delta V^{\ddagger}$ consistent with a dissociatively activated reaction.

In addition, increased steric crowding of the reaction centre by increasingly

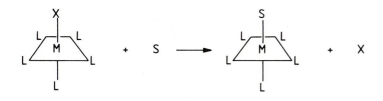

Fig. 4.2

more bulky coligands leads to an increase in the reaction rate, consistent with the dissociative mechanism. Thus the rate of aquation (displacement of a ligand by a water molecule) of the complexes, $[Co(NH_2R)_5Cl]^{2+}$, progressively increases as R is varied from R = H to R = Bu^i.

Influence of the leaving group

In these dissociatively activated reactions the nature of the leaving group is of major importance in defining the rate of the reaction. In general, the propensity of the leaving group to depart, correlates with the metal-ligand bond strength, which in its turn is related to the type of site. In relatively crude terms if the metal centre is 'hard' as in $[Co(NH_3)_5X]^{n+}$ then the harder ligands will dissociate less readily than the softer ligands. We observe the trend in the first order rate constants: I > Br > Cl > F. In contrast, if the reaction centre is 'softer' (by introducing electron-releasing ligands around the same metal in the same oxidation state) as in $[Co(CN)_5X]^{n+}$ then the 'hard' ligands tend to dissociate more readily, F > Cl > Br > I.

Influence of the spectator ligands

Any substitution reaction is sensitive to the nature of the other ligands around the metal centre, since it is these ligands which help define the relative electron-richness of the metal site and hence its substitution characteristics. However, when compared to square-planar complexes, the reactions of octahedral compounds show a remarkable insensitivity to the nature of the spectator ligands. Thus there is no dominant effect from the *trans*-ligand as we saw in square-planar substitution; only hydrido-, nitrido- and sulfito-ligands come close to operating a significant, and general, *trans* labilizing effect.

Again the most comprehensive collection of data concerning the electronic influences of various spectator ligands is that on the cobalt(III) complexes, *cis*- or *trans*-$[Coen_2ACl]^{n+}$ (en = $NH_2CH_2CH_2NH_2$). Two types of labilizing effects are realized in these systems:

(1) π-electron donors labilise if they are *cis* to the leaving group, by donating electron-density into a metal orbital used by the leaving group and in effect pushing the leaving group out as illustrated in Fig. 4.3;

(2) σ-electron donors or π-electron acceptors labilise if they are *trans* to the leaving group, by weakening the Co-Cl bond.

Fig. 4.3

Stereochemistry of the substitution

In general the substitution reactions at an octahedral site occur with complete retention of stereochemistry and, as shown in Fig. 4.4, this is consistent with a square-based pyramidal five-coordinate intermediate.

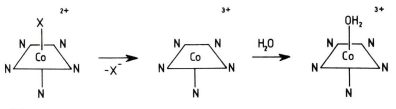

Fig. 4.4

However the energy difference between this intermediate and the alternative trigonal bipyramidal form must be very small, since for the complexes $[CoL_4ACl]^{n+}$ ($L = NH_3$ or $\frac{1}{2} H_2NCH_2CH_2NH_2$), if the group A is capable of π-electron donation then stereochemical change is observed giving a mixture of *cis* and *trans* isomers. This stereochemical change seems most likely to be a consequence of the π-donation being most effective when the group A is contained within the trigonal plane of a trigonal bipyramidal intermediate. Once formed, this trigonal bipyramidal intermediate can then be attacked by the solvent nucleophile. The attack occurs in the trigonal plane of the intermediate, and on purely statistical grounds we would expect that 66.6 per cent of the product would be stereochemically *cis* (attack adjacent to the group A) and 33.3 per cent of the product would be stereochemically *trans* (attack opposite the group A). In practice, of course, these statistical values are not observed, at least in part because the sites of attack in the trigonal bipyramid are subject to various electronic and steric factors which leads to discrimination by the nucleophile as to which position is attacked.

4.2   Displacement of coordinated solvent

Solvent exchange reactions

Having replaced the leaving group by a solvent molecule, we now look at the pathways by which the solvento-ligand is replaced by another molecule. One of the simplest reactions, at least conceptually, is the solvent exchange process shown in Fig. 4.5. This type of reaction has been studied for a relatively wide range of transition metal ions in oxidation state II and III. In particular we are able to discuss the effect on the intimate mechanism of the substitution reaction of moving along the first row of the transition metals whilst keeping all the following constant:

(1)   the oxidation state of the metal;
(2)   the geometry of the metal;
(3)   the leaving group;
(4)   the entering group.

Hence the influence of size, charge and electron configuration on the rate and mechanism can be described.

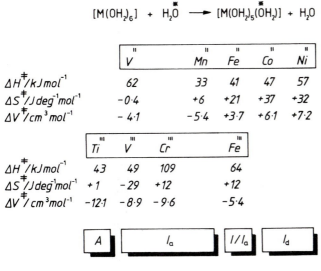

$$[M(OH_2)_6] + H_2\overset{*}{O} \longrightarrow [M(OH_2)_5(O\overset{*}{H_2})] + H_2O$$

	V^{II}		Mn^{II}	Fe^{II}	Co^{II}	Ni^{II}
$\Delta H^{\ddagger}/kJ\,mol^{-1}$	62		33	41	47	57
$\Delta S^{\ddagger}/J\,deg^{-1}mol^{-1}$	-0.4		$+6$	$+21$	$+37$	$+32$
$\Delta V^{\ddagger}/cm^3\,mol^{-1}$	-4.1		-5.4	$+3.7$	$+6.1$	$+7.2$

	Ti^{III}	V^{III}	Cr^{III}	Fe^{III}
$\Delta H^{\ddagger}/kJ\,mol^{-1}$	43	49	109	64
$\Delta S^{\ddagger}/J\,deg^{-1}mol^{-1}$	$+1$	-29	$+12$	$+12$
$\Delta V^{\ddagger}/cm^3\,mol^{-1}$	-12.1	-8.9	-9.6	-5.4

A		I_a		I/I_a	I_d

Fig. 4.5

The experimental problem with this type of this study (as in the previous section) is that the solvent is also the entering group and hence the determination of the rate law of the reaction reveals little about the mechanism. Consequently, our interpretation of the mechanism of the reactions is based on the extra kinetic parameters, primarily $\Delta S^{\ddagger}$ and $\Delta V^{\ddagger}$.

From the data shown in Fig. 4.5 it is clear that upon moving from left to right across the first row of the transition metals there is a change from an associative to a dissociative mechanism irrespective of the oxidation state of the metal. Why this should be, is relatively easy to rationalize considering the approach of a water molecule to an octahedral complex, as shown in Fig. 4.6.

For a six-coordinate metal centre, the direction of attack of the water molecule must be along one of the C_3 axes, along which the t_{2g} orbitals point. Consequently if these orbitals are not occupied by electrons then they are ideally suited to accept the pair of electrons from the incoming nucleophile. However, as we proceed across the first period of the transition metals these orbitals become progressively filled with electrons and the now full (or partially filled) t_{2g} orbitals will repel the approaching water molecule, leaving the complex little option but to accomplish the substitution reaction by a dissociative pathway.

Inspection of the first order rate constants for these exchange reactions shows that a wide range of labilities is observed. For many years these rates, and in particular the kinetic robustness of $d^{3,}$ typified by Cr(III), and low spin $d^{6,}$ typified by Co(III), were rationalized in terms of the crystal-field stabilization energy associated with these electron configurations in their substitution reactions. Briefly, this stabilization energy is calculated assuming that each electron in a t_{2g} orbital is stabilized by $\frac{2}{5}\Delta_0$ and each

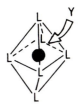

Fig. 4.6

electron is an e_g orbital destabilized by $\frac{3}{5}\Delta_0$ (where Δ_0 is the energy separation between the t_{2g} and e_g orbitals in an octahedral field). Thus for high spin Cr(II), a $t_{2g}^3 e_g^1$ ion, the crystal-field stabilization energy is $3(\frac{2}{5}\Delta_0) + (-\frac{3}{5}\Delta_0) = \frac{3}{5}\Delta_0$. A similar calculation is then performed on the putative intermediate structures (square-based pyramid or trigonal bipyramid for dissociative mechanisms, and pentagonal bipyramid or octahedral wedge for associative mechanisms). The difference in the stabilization energies for the octahedral and intermediate structures is the crystal-field activation energy (CFAE), which is a contributor to the total activation energy of the system. This calculation is particularly successful in predicting that the CFAE is at a maximum at d^3, low spin d^6 and d^8 ions, which consequently are the most robust species kinetically. This rationalization may be largely fortuitous since CFAE contributes only a small part to the total activation energy and there are many other factors which control whether a site reacts rapidly or slowly. However it must be remembered that a full or half full-shell of electrons imparts a degree of kinetic robustness on the metal ion. Thus we would expect to see relatively low rates of substitution for metal ions with d^3, d^5 (high spin) and d^6 (low spin) configurations.

Substitution of coordinated solvent

Having considered the simple exchange reaction of the first row transition metal ions, we next consider what happens to the reactivity pattern when the nature of the entering group is changed. In essence, we would expect only relatively minor changes to the reactivity pattern described above. Any changes would reflect the varying nucleophilicity of the entering group. A large number of studies on a wide range of complexes with electron configurations d^0 to d^{10} and in a wide variety of coordination numbers has shown that the intimate mechanism of substitution of the metal ions is the Eigen–Wilkins mechanism shown in Fig. 4.7.

$$[ML_6]^+ \; + \; Y^- \; \rightleftharpoons \; [ML_6]^+.Y^- \; \longrightarrow \; [ML_5Y] \; + \; L$$

Fig. 4.7

In this mechanism the initial step involves an outer sphere association of the nucleophile with the metal complex; that is the nucleophile enters the solvation sphere of the complex. The major force holding these two species in close proximity at this stage is often predominantly coulombic, reflecting their different charges, in which case this outer sphere association is correctly considered as an ion-pair. When either species is uncharged, or both species have the same charge, then hydrogen bonding or much weaker van der Waal's forces bind the two together. Once in the solvation sphere, the nucleophile can either attack the metal centre if the electron configuration of the metal permits it, or must await the dissociation of one of the coordinated water molecules before it can then enter the coordination sphere of the metal ion. If the nucleophile is on the side remote from the dissociating water molecule, it will stand little chance of being able to bind at the vacant site before another, much closer water molecule in the solvation sphere binds. The nucleophile must await the dissociation of a water molecule in close proximity to itself. The important point about any dissociatively activated substitution reaction involving the replacement of coordinated water is that not all acts of dissociation of the coordinated solvent lead to a substitution.

This last point is further amplified when we look at the replacement of coordinated water in the reactions of the substitutionally robust $[Co(NH_3)_5(OH_2)]^{3+}$, shown in Fig. 4.8.

$$[Co(NH_3)_5(OH_2)]^{3+} + Y^{2-} \overset{K_o}{\rightleftharpoons} [Co(NH_3)_5(OH_2)]^{3+}.Y^{2-} \overset{k}{\longrightarrow} [Co(NH_3)_5Y]^{+}$$

$$k_{obs} = \frac{kK_o[Y^{2-}]}{1 + K_o[Y^{2-}]}$$

Fig. 4.8

In this system, when $Y^{2-} = SO_4^{2-}$, deviations from a simple first order dependence on the concentration of the nucleophile are observed at relatively high concentrations of sulfate ion. This deviation from simple kinetics is attributable to a buildup of a significant concentration of the ion-pair, because of the strong electrostatic interaction between the 3+ cation and the 2– anion. Because of this deviation, it is possible to determine the values of K_o and k separately. Studies with a wide range of nucleophiles and a similar kinetic analysis (or estimations of the value of K_o using an electrostatic approach) shows that the value of k (the interchange process) is independent of the nature of the nucleophile as would be expected for an I_d mechanism. However, the value of k is about one-sixth the value for the water exchange rate constant established for $[Co(OH_2)_6]^{3+}$. This has been interpreted in terms of the outer sphere complex with the nucleophile in the solvation sphere jostling for position with the water molecules, and only once in every six dissociations of the coordinated water molecule does the nucleophile manage to beat the water molecules in the solvation sphere to the vacant site.

This relationship between the value of the interchange rate constant, k, and the exchange rate constant, k_{ex}, ($k_{ex} / k \simeq 6$), for the reaction of $[Co(NH_3)_5(OH_2)]^{3+}$ should not be looked on as simply reflecting the relative numbers of dissociable water molecules in this species and $[Co(OH_2)_6]^{3+}$. There are many other factors operating, not least of which are the relative nucleophilicities of Y and water for the particular metal site, and the differing steric requirements of the various nucleophiles and water. Even for the generalized series of complexes, $[M(NH_3)_5OH_2]^{3+}$, the ratio k_{ex} / k varies dramatically from a value of about 30 (M = Cr) to about 0.1 (M = Rh).

The important feature about the kinetics of the replacement of coordinated water, as exemplified by the system shown in Fig. 4.8, is that the first order dependence on the concentration of the nucleophile is not ascribable to associative, bimolecular attack on the metal centre, but is attributable to the formation of the outer sphere complex. The mechanism of the reaction is best described as I_d.

The use of substitutionally robust complexes to study the rates of solvent exchange has several advantages over the more labile complexes. For instance, the nature of the coligands can be varied and hence their steric and electronic influences on the mechanism can be investigated. Most importantly we can study the stereochemistry of the reactions, an impossibility in the $[M(OH_2)_6]^{n+}$ systems without some rather elaborate isotopic labelling in the more kinetically robust systems. One other advantage is that the reactions of the isolated solvento-complex can be studied in a different, non-interacting solvent. This allows us to look at the influence of the released solvent molecule on the kinetics, as shown in the example in Fig. 4.9.

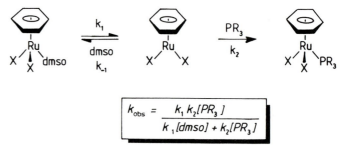

$$k_{obs} = \frac{k_1 k_2 [PR_3]}{k_{-1}[dmso] + k_2[PR_3]}$$

Fig. 4.9

In this study, the substitution of the coordinated dmso (dmso = Me_2SO) by PPh_3 is studied in the non-coordinating 1,2-dichloroethane, and the mechanism has been shown to involve rate-limiting dissociation of dmso because of the form of the rate law: the addition of free dmso inhibits the reaction and at high concentrations of PPh_3 the rate of the reaction becomes independent of the concentration of the nucleophile. The similarity in the kinetic form for this reaction and that of the reaction shown in Fig. 4.8 is clear but the origin of the saturation behaviour in the two systems is very different. In the reactions with $[Co(NH_3)_5(OH_2)]^{3+}$ the limiting zero order dependence on the concentration of the nucleophile is due to the buildup of appreciable concentrations of the outer sphere complex. In contrast there is no accumulation of intermediates in the system shown in Fig. 4.9. The deviation from first-order dependence on the concentration of PPh_3 occurs only when $k_2[PPh_3] \gg k_{-1}[dmso]$.

In the solvent exchange reactions of $[MH_2S_2(PPh_3)_2]$ (S = MeCN or Me_2CO, M = Rh or Ir), a dissociative mechanism is indicated by $\Delta S^{\ddagger} = $ 10–30 J K^{-1} mol^{-1}. In this case there is a strong *trans*-labilizing force of the hydrido-ligands which favours the dissociative mechanism.

Finally, in this section, the replacement of the tetrahydrofuran (thf) ligands in $[M(CO)_5(thf)]$ (M = Cr, Mo or W) by piperidine, $P(OEt)_3$ or PPh_3, has been studied by measuring $\Delta V^{\ddagger}$ of the reactions. This study hows that as we go down Group VI the reaction becomes increasingly more associative in nature. This is a trend we will see again in other reactions and is attributable, at least in part, to the increased size of the metal atom as we go from the first row to the third row metal.

4.3   Substitution by direct replacement

Associative and dissociative pathways

The preceding section of this chapter describes the details of how octahedral complexes undergo substitution involving the intermediacy of a solvento-species. The description that results is highly detailed because of the use of kinetically robust complexes, permitting the independent isolation and mechanistic characterization of each of the two stages in these reactions. However, it is clear that there is nothing special about these robust complexes, and that analogous pathways also operate in the reactions of the more labile complexes. Also, we have seen that complexes containing metals in both high and low oxidation states use similar mechanisms.

The topic of this section is direct substitution reactions and involves a wide variety of reactions, covering a multitude of complexes and nucleophiles. The approach that can be taken in this area is, as a consequence, not so rigorous. Nevertheless, we will see the same general characteristics: a preponderance of dissociative activation pathways, especially for first row transition metal compounds, but an increasing amount of information showing the operation of associative pathways. For instance, in the reactions of $[MnX(CO)_5]$ (X = halide) with molecules such as 1,10-phenanthroline to form $[MnX(CO)_3(phen)]$ the positive values of $\Delta S^{\ddagger} = 37 J$ K^{-1} mol^{-1} and $\Delta V^{\ddagger} = 20.6$ cm^3 mol^{-1} demonstrate the dissociative nature of the these reactions. However in the reactions of $[Mn(\eta^5\text{-}C_5H_5)(CO)_3]$ with phosphines, the first order dependence on the concentration of the nucleophile together with the negative values of $\Delta S^{\ddagger}$ strongly implicate an associative mechanism.

In a further example of associative mechanisms, the displacement of carbon monoxide ligands in $[V(CO)_5(NO)]$ by phosphines occurs by two pathways: a dissociative route (independent of the concentration and nature of the phosphine) and an associative route which shows a first order dependence on the concentration of the nucleophile. Hence for a six-coordinate, closed shell, vanadium(0) complex, both associative and dissociative activation are of comparable energy. Indeed this sort of behaviour is by no means restricted to first row transition metal elements and although the reaction of PPh_3 with $[WX(CO)_4(NO)]$ (X = Cl, Br, or I) occurs by a purely dissociative pathway, the reaction with the more nucleophilic PBu_3^n occurs by a mixture of dissociative and associative pathways. The latter route is characterized by a first order dependence on the concentration of PBu_3^n and $\Delta S^{\ddagger} = -87.4 JK^{-1}mol^{-1}$.

Why then do these coordinatively saturated complexes have available both an associative and a dissociative pathway for substitution? The only other cases of an associative pathway in octahedral complexes that we have met so far are the reactions of the early transition metals containing empty t_{2g} orbitals: a situation clearly not pertaining here. In order to render the associative mechanism energetically viable in closed shell complexes there must be a way in which these species can attain coordinative unsaturation, without entirely dissociating a ligand. Inspection of the coligands in the complexes that undergo these associative pathways reveals that they all have the common feature that they can donate varying numbers of electrons to the metal. The range of ligands that can do this is really quite large and just a few of the more common ones are shown in Fig. 4.10.

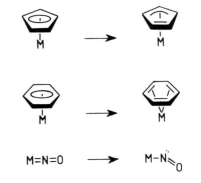

Fig. 4.10

In our original example of the two manganese complexes we see that $[MnX(CO)_5]$ contains none of the ligands illustrated in Fig. 4.10 and thus a dissociative mechanism is enforced. In contrast, $[Mn(\eta^5\text{-}C_5H_5)(CO)_3]$ contains the pentahapto-cyclopentadienyl ligand and upon approach of a nucleophile towards the manganese, the cyclopentadienyl-ring can change its bonding to the metal to become a trihapto-ligand. In this way the metal centre relieves itself of the burden of the extra two electrons donated by the nucleophile. In effect the metal pushes two electrons onto the cyclopentadienyl-ligand without having to completely dissociate any ligand, and hence retains its closed shell configuration. In a similar manner we see that in the complexes containing nitrosyl ligands the attack of a nucleophile at the metal can trigger a change in bonding of the nitrosyl from a linear, three electron donor to a bent, one electron donor, exactly the same effect as we saw in the substitution reactions of tetrahedral sites in Chapter 3. Once the leaving group has departed, the cyclopentadienyl- or the nitrosyl-ligand can revert to its original form of bonding to the metal in order to retain the closed shell configuration of the product.

Evidence for the shift mechanism

Direct evidence for a cyclopentadienyl-ligand changing its bonding during associative substitution pathways comes from studies on the rhenium system shown in Fig. 4.11.

In this reaction the substitution of the carbon monoxide ligand by PMe_3

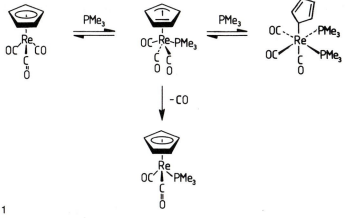

Fig. 4.11

shows a first order dependence on the concentration of the nucleophile and thus implicates the intermediate containing a trihapto-cyclopentadienyl residue. In the presence of an excess of PMe_3 this intermediate can be trapped, isolated and characterized as the bis(phosphine) complex which necessarily contains the monohapto-cyclopentadienyl-ligand: This demonstrates the rapid response and flexibility of the cyclopentadienyl-ligand to changes in the coordination sphere of the metal.

4.4 More general considerations: other coordination numbers

The activation of systems towards associative mechanisms by some ligands as described above is not restricted to six-coordinate complexes of the reactants, but to the presence of ligands on a coordinatively saturated centre.

In the five-coordinate complexes of Group IX, $[M(\eta^5\text{-}C_5Me_5)(CO)_2]$ (M = Co or Rh) substitution of the carbonyl ligands occurs by an associative mechanism. In this example we see a group trend in that the rhodium complex reacts faster than the cobalt analogue, which can be rationalized in terms of the larger size of the rhodium atom and the weaker binding of the CO ligand to the rhodium centre. However there is a lesson to be learnt from studies on this type of system: just the presence of a ligand capable of changing the number of electrons that it formally donates to the metal does not mean that the complex will always adopt an associative mechanism. In the reaction of the analogous $[Co(\eta^5\text{-}C_5H_5)(PPh_3)_2]$ with PMe_3 the pathway is exclusively dissociative.

In the nine-coordinate complexes, $[M(\eta^5\text{-}Ind)Cl(CO)_3]$ (M = Mo or W, Ind = C_9H_7) substitution of a carbonyl ligand by PPh_3 occurs by a mixture of dissociative and associative pathways. Again, we see the trend we have observed before in other systems, that the heavier element reacts more rapidly. In this case this greater lability of the tungsten analogue is entirely due to a faster associative pathway, and the dissociative pathway is faster for the molybdenum species.

In the complexes, $[M(\eta^5\text{-}C_5H_5)(CO)_3(\text{tht})]$ (M = V, Nb or Ta; tht = C_4H_8S) substitution of the tetrahydrothiophen ligand by phosphines occurs by a dissociative pathway for vanadium but for the niobium and tantalum analogues, with their larger metal centres, an associative route is observed.

Finally, in eight-coordinate complexes, $[M(\eta^5\text{-}C_5Me_5)_2(CO)_2]$ (M = Ti, Zr or Hf), displacement of a carbonyl ligand by a phosphine shows the same trend as we observed for the Group V metals above, namely, the first row metal reacts by a dissociative pathway exclusively whereas the heavier members of the group react, at least in part, by an associative route.

Substitution at seventeen and nineteen electron species

We have seen how an associative pathway in coordinatively saturated complexes can proceed via an intermediate which retains the closed shell electron configuration. But what happens if the metal cannot attain the closed shell configuration in the intermediate? This situation is met in the relatively rare cases when the reactant complex has formally either a seventeen or nineteen electron count.

In the reactions of $[V(CO)_6]$ or $[W(\eta^5\text{-}C_5H_5)(CO)_3]$ with PR_3 in which the phosphine replaces a coordinated carbonyl, the mechanism of the reactions is associative, as indicated by the first order dependence on the concentration of phosphine and the large negative values of $\Delta S^{\ddagger}$. Thus the pathway involving a 19 electron intermediate is energetically more favourable than the dissociative route involving a 15 electron species. It has been proposed that the associative pathway in the reactions of 17 electron species is favoured because of the interaction of the nucleophile with the singly occupied orbital on the metal (a two-centre, three electron interaction), or that the odd electron is delocalized onto a ligand. In the reactions of the seven-coordinate, 17-electron species, $[V(\eta^5\text{-}C_5Me_5)_2(CO)]$, again we see an associative mechanism for the CO exchange reaction associated with $\Delta H^{\ddagger} = 37.4$ kJ mol^{-1} and $\Delta S^{\ddagger} = 88$ J K^{-1} mol^{-1}. The factors which decide whether the mechanism of a substitution reaction operates by an associative or a dissociative pathway seem to remain the same irrespective of coordination number.

There are very few 19 electron species which have been the subject of mechanistic studies. The studies that have been performed show the same general characteristic: that the 19 electron species must convert to a 17 electron species either intramolecularly or by employing a dissociative pathway. Thus the six-coordinate 19 electron species, $[Fe(\eta^5\text{-}C_5H_5)(\eta^6\text{-}arene)]$, has to rearrange to $[Fe(\eta^5\text{-}C_5H_5)(\eta^4\text{-}arene)]$ before it is susceptible to attack by PR_3 which subsequently displaces the arene. The five-coordinate complex shown in Fig. 4.12 is better considered as an 18 electron species with the odd electron delocalized on the phosphine ligand. Here the substitution of one of the CO ligands occurs by a dissociative mechanism with $\Delta H^{\ddagger} = 100$ kJ mol^{-1}, $\Delta S^{\ddagger} = 47$ J K^{-1} mol^{-1}. In both cases the systems have a prohibitively slow associative pathway, whether it contains a formal 21 electron configuration, or a formal 20 electron count at the metal and the odd electron on a ligand.

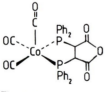

Fig. 4.12

5 Catalysed substitution reactions

In the previous chapter, substitution reactions at a variety of metal sites were discussed. In this chapter we will see how the rates of these substitution processes can be accelerated by base, acid or electron transfer.

5.1 Base catalysed hydrolysis

The most extensively studied system showing base catalysis is that of the Werner-type complexes of cobalt(III), chromium(III), ruthenium(III), rhodium(III), and iridium(III). Again the majority of the details have been established for the cobalt systems and thus we will concentrate exclusively on the reactions of this element. The initial interest in these reactions was the apparent anomaly that the reaction of hydroxide, such as that shown in Fig. 5.1, appeared to be an exception to the general rule that the rate of substitution at cobalt(III) complexes showed the characteristics of a D mechanism and was independent of the nature and concentration of the nucleophile.

$$[CoCl(NH_3)_5]^{2+} + HO^- \longrightarrow [Co(OH)(NH_3)_5]^{2+} + Cl^-$$

Fig. 5.1

Also, the reactions with hydroxide were in general several orders of magnitude faster than the corresponding rates of solvolysis typically studied in acidic solution. There was much discussion, mostly in the 1950s and 1960s, concerning the mechanism of this reaction but it is now established beyond doubt that the mechanism is that first described in 1937 by Garrick, which in turn was based on an analogous proposal as early as 1907 by Bjerrum. The now accepted mechanism is that shown in Fig. 5.2.

The key feature of this mechanism is the deprotonation of an amine ligand to generate the conjugate base of the complex, which then undergoes dissociation of the leaving group to generate a five-coordinate intermediate. Subsequent attack by water on the five-coordinate, deprotonated intermediate yields the product. The dissociative nature of this mechanism (often the dissociation step is rate-limiting), and the essential formation of the conjugate base has led to this mechanism being labelled D_{cb} (dissociative conjugate base). The evidence that favours this

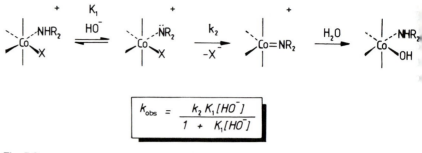

$$k_{obs} = \frac{k_2 K_1 [HO^-]}{1 + K_1 [HO^-]}$$

Fig. 5.2

mechanism comes from several sources and we shall discuss each separately.

Presence of acidic protons

Clearly the presence of easily abstracted protons is essential for the D_{cb} mechanism to operate. Complexes that are devoid of such protons, e.g. $[Co(CN)_5Cl]^{3-}$ and *trans*-$[CoCl_2(Ph_2PCH_2CH_2PPh_2)_2]^+$, do not undergo a base catalysed reaction.

Another requirement for the D_{cb} mechanism is that the protons on the ligands are easily abstracted. In the majority of cases the rate of proton exchange is usually close to the diffusion-controlled limit and is much faster than the dissociation of the leaving group. Thus for $[CoCl(NH_3)_5]^{2+}$, $k_{-1}/k_2 = 7-100 \times 10^5$. However, under certain circumstances proton exchange can be rate-limiting; most particularly for the complexes shown in Fig. 5.3.

This demonstration that the act of deprotonation (shown by measuring the rate of proton exchange or investigating the general base catalysis of the reaction) represents one of the best proofs of the D_{cb} mechanism.

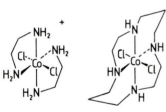

Fig. 5.3

Dissociative mechanism

The dissociative activation of the vast majority of cobalt(III) complexes undergoing the base hydrolysis mechanism is substantiated by the values $\Delta S^{\ddagger} = +80$ to $+160$ J K^{-1} mol^{-1} and $\Delta V^{\ddagger} = +19$ to $+43$ cm^3 mol^{-1}. However, the multistep nature of the mechanism means that these parameters relate to the accumulation of several elementary processes making the interpretation of these values ambiguous. Somewhat surprisingly, the values of $\Delta V^{\ddagger}$ are independent of whether dissociation of the leaving group or deprotonation of a coordinated amine is rate-limiting.

Increasing the steric bulk on the complex increases the rate of base hydrolysis, as would be expected if the reaction is dissociatively activated. Thus the reactions of $[CoCl(NH_2Me)_5]^{2+}$ or $[CoCl(NH_2CHMeEt)_5]^{2+}$ with hydroxide are 10^4 to 10^5 times faster than that of $[CoCl(NH_3)_5]^{2+}$. This increased lability is not due to a change in the rate of deprotonation or acidity of the amine.

The best evidence that the reaction is dissociatively activated, would come from the demonstration that a five-coordinate intermediate was generated in the base hydrolysis reaction. So far, no five-coordinate intermediate has accumulated in sufficient concentrations to be detected directly by spectroscopic methods, although it is possible to demonstrate the existence of an intermediate of reduced coordination number by competition experiments, such as those illustrated in Fig. 5.4.

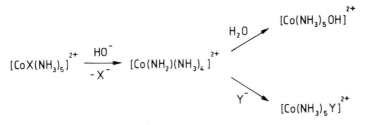

Fig. 5.4

Studies on this system rely on the large difference in sensitivity of the complexes to base hydrolysis depending upon the nature of the leaving group. Thus when the leaving group is chloride, bromide, etc., the base hydrolysis reaction is very rapid, whereas the azide or thiocyanate analogues react very sluggishly. Hence the azide and thiocyanate ions can be used as 'competition ions' during the base hydrolysis of the chloro- or bromo-complexes.

Such competition studies have demonstrated the following.

1. The competition ratio, $[CoY(NH_3)_5]^{2+} / [Co(OH)(NH_3)_5]^{2+}$ is linearly dependent on the concentration of Y^- as expected from the simple consideration that the more Y^- present, the more five-coordinate intermediate will be 'trapped' by it.

2. The competition ratio is independent of the concentration of hydroxide ion, as expected since the role of the hydroxide is purely catalytic.

3. The least solvated end of the nucleophile, Y^-, is captured by the five-coordinate intermediate. Thus for NCS^- it is the sulfur end which is initially captured by the five-coordinate cobalt atom. Of course the 'hard' cobalt centre dislikes being bound to the 'soft' sulfur and relatively rapidly, the system isomerizes to the thermodynamically more stable N–bound isomer.

A key demonstration of the existence of a five-coordinate intermediate is that the competition ratio should be independent of the nature of X, the leaving group. However this is not strictly observed in the cobalt systems: there is a slight variation in the competition ratio depending on the leaving group. A five-coordinate intermediate will only behave independently of its origins, if it has time to equilibrate with its surroundings prior to capture. This means, not only dissociating the leaving group from the metal, but also for the leaving group to have time to leave the solvation shell. If the leaving group is still present in the solvation shell at the time of the intermediate's capture, then it may influence the outcome of the competition between water and Y^-; in effect it acts as a 'memory' for the intermediate.

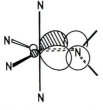

Fig. 5.5

Reactivity of the conjugate base

The unusually rapid reactivity of cobalt(III) complexes in the base hydrolysis reaction must be ascribed to the extreme lability of the conjugate base. The conjugate base would be expected to be more labile than its conjugate acid purely on charge grounds, since it has one less positive charge. The leaving group is often anionic, and so charge separation problems in the transition state of the dissociation are less for the conjugate base than they are for the parent complex. This effect alone cannot account for the very large differences in reactivity of the acid–base pair. The major labilizing force in the reaction has been proposed to be the formation of a π-bond, by overlap of the lone pair of electrons on the deprotonated amine ligand with the metal as shown in Fig. 5.5.

In effect, the overlap of the lone pair of electrons with the metal orbitals pushes the leaving group out, whilst retaining a formal 18 electron configuration at the cobalt.

That the lone pair of electrons on the amido-ligand overlaps with the orbitals on the metal has been shown in the system shown in Fig. 5.6.

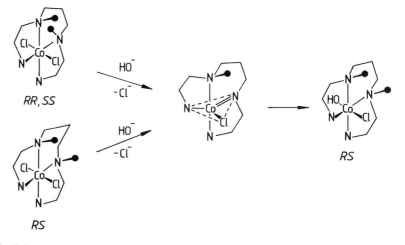

Fig. 5.6

The two dichloro-complexes differ only in the orientation of the secondary amine protons. As a direct consequence of the act of base hydrolysis both complexes give rise to the same *RS*-diastereoisomeric product. In order that this can have happened with the *RR,SS*-diastereoisomer, the secondary nitrogen atom must have become planar during the reaction.

The formation of a π-bond is a strongly labilizing force and this has been illustrated with systems of the type shown in Fig. 5.7.

Here, the use of linear quadridentate ligands such as $O_2CCH_2NHCH_2CH_2NHCH_2CO_2$ gives rise to species with very different π-bonding capabilities at the secondary amine donors (the only acidic protons in the complex). With the α-*cis* geometry (top) the constraints of the quadridentate ligand preclude effective overlap of the lone pair of

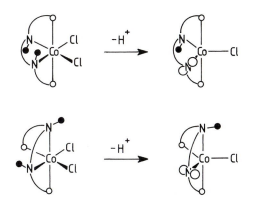

Fig. 5.7

electrons with the metal orbitals, whereas with the ß-*cis* geometry (bottom) effective overlap is achieved. Consequently the ß-*cis* complex reacts over 10^4 times more rapidly than the α-*cis* isomer. The rates of aquation of the α-*cis* and ß-*cis* isomers are essentially the same.

Stereochemistry of base hydrolysis reactions

In contrast to the solvolysis reactions, the base hydrolysis reactions of these cobalt(III) complexes occur with extensive rearrangement. However, accurate product analyses show that the product distribution is independent of the leaving group and the isomeric form of the starting material. For instance, for the complexes, *cis*- or *trans*-[Co(OH)X(en)$_2$]$^+$ (X = Cl or Br, en = NH$_2$CH$_2$CH$_2$NH$_2$), shown in Fig. 5.8, the distribution of isomeric products, [Co(OH)$_2$(en)$_2$]$^+$, was the same: *cis* = 7 ± 3 per cent, *trans* = 94 ± 3 per cent. This behaviour is not only consistent with a dissociative mechanism, but is also indicative of a trigonal bipyramidal intermediate.

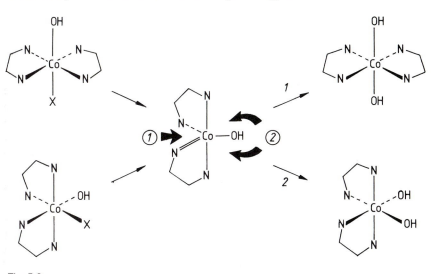

Fig. 5.8

Use of the isotopically labelled *trans*-[CoX(^{15}NH$_3$)(NH$_3$)$_4$]$^{2+}$ gives a particularly interesting stereochemical result, as shown in Fig. 5.9.

Use of this isotopic label in such a simple complex, permits determination of the stereochemical consequences of base hydrolysis reaction free from complications such as, stereochemical influences of various coligands, presence of hydrophobic residues on coligands and specific, or localized solvation effects, all of which may introduce a distortion into the stereochemistry of the reaction.

The observed product distribution, *cis*/*trans* = 1.0, is independent of the nature of X and can be rationalized in terms of a trigonal bipyramidal intermediate in which the site of deprotonation is *cis* to the leaving group. However there is a restriction on the position of attack of the water on the five-coordinate intermediate, and that is it must attack at the same relative position from which X has left. For an intermediate that has had time to equilibrate with its surroundings, this means attack at the trigonal plane at a site adjacent to the π-bond and would give rise to the observed product distribution, as illustrated in Fig. 5.9. This study also indicates that it is a *cis* ammonia ligand which generates the active amido-group.

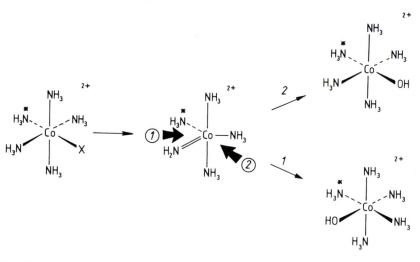

Fig. 5.9

Base hydrolysis of imido-complexes

A system that has many similarities to the classic base hydrolysis mechanism, is the base catalysed substitution of halide for methoxide in the reactions of *trans*-[M(NH)X(Ph$_2$PCH$_2$CH$_2$PPh$_2$)$_2$]$^+$ (M = Mo or W; X = Cl, Br or I) shown in Fig. 5.10.

The mechanism consists of the rapid deprotonation of the imido-ligand to generate the nitrido-species. Notice here that the nitrido-group is *trans* to the leaving group but the nitrido-ligand operates such a strong *trans* labilizing effect that the halide rapidly dissociates. This dominant *trans* effect is rare in substitution at octahedral sites as we have pointed out before, and is more usual in substitution at square planar geometries.

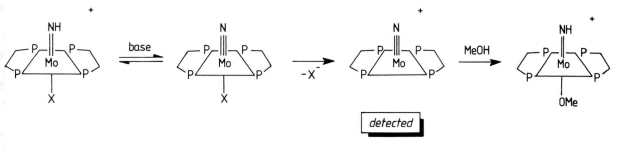

Fig. 5.10

Once formed, the derived five-coordinate intermediate can now be attacked by methoxide. The replacement of a halide by methoxy group renders the nitrido ligand sufficiently basic that it abstracts a proton from the solvent, methanol, to form the imido-product. This change in the basicity of a ligand as a consequence of other changes in the coordination sphere is a general phenomenon which we shall meet again in Chapter 8. The important and rare aspect of this reaction is that the five-coordinate, square-based pyramidal intermediate is sufficiently long-lived, and attains sufficiently high concentrations, that it can be detected and characterized spectroscopically. Its spectral characteristics and reactivity are independent of its origins, entirely consistent with a D mechanism. To be able to monitor the formation and decomposition of a five-coordinate intermediate is not common, since the system has to fulfil some fairly stringent requirements. Not only must the intermediate be formed rapidly, but once formed it must react subsequently relatively slowly; this is often difficult to accomplish in the presence of a coordinating solvent. The success of the *trans*-$[M(NH)X(Ph_2PCH_2CH_2PPh_2)_2]^+$ is that the five-coordinate intermediate is impervious to attack by methanol, and reacts only with low concentrations of methoxide ion.

5.2 Acid catalysed hydrolysis

The influence of acid on the outcome and lability of transition metal complexes is a wide area. The interaction of protons with transition metal complexes can be classified into three types:
(1) protonation of a ligand which then dissociates from the metal;
(2) protonation of a ligand which then labilizes another ligand to dissociation;
(3) protonation at the metal labilizing the substitution of a ligand.

We shall discuss each of these areas in turn. However, we will restrict our discussion to truly catalytic roles for protons, that, is in which there is no nett consumption of protons. The general types of reactions also extend to reactions which are not catalytic in protons.

Protonation of the leaving group

Protonation of a ligand can facilitate the dissociation of that ligand, particularly when the ligand in question is the conjugate base of a relatively weak acid. The only requirement of the ligand is the availability of a lone pair of electrons to which the proton can bind. Thus for complexes of the type, *trans*-$[CoX_2en_2]^+$ (X = F, Cl, Br or I, en = $NH_2CH_2CH_2NH_2$), the rate of substitution of the first halide in acidic aqueous media is independent of the acid concentration for X = Cl, Br or I since each of the corresponding acids are strong acids in this solvent. However for X = F the substitution is strongly acid catalysed because of the much weaker acidity of HF in water. The mechanism of the reaction is shown in Fig. 5.11.

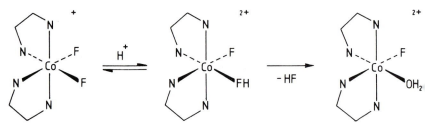

Fig. 5.11

Protonation of the fluoro-group changes the nature of the leaving group from an anionic species to the neutral HF molecule. This will help the dissociation of the residue since now in the transition state of the dissociation problems of charge separation are not so pronounced: the cationic metal fragment releases the neutral residue from its coordination sphere more readily than it releases an anion.

This principle of protonating a leaving group is manifest in the non-catalytic acid catalysed hydrolysis of various coordinated oxyanions, particularly at cobalt(III) centres. Probably the most well studied system is the hydrolysis of carbonato-complexes such as $[Co(O_2CO)(NH_3)_5]^+$. The mechanism of the acid-dependent pathway is shown in Fig. 5.12.

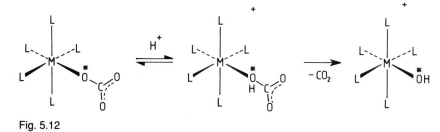

Fig. 5.12

In this system, rapid substitution is observed not only because of the susceptibility of the system to protonation but also because the subsequent cleavage step does not involve any cleavage of a cobalt–oxygen bond, as demonstrated by the fact that the isotopically labelled oxygen on the carbonato-ligand is retained in the product.

Another example of protonation of a leaving group is the acid catalysed dissociation of chelate rings in multidentate ligands, as shown by the example in Fig. 5.13, of the reactions with *trans-*[FeCl$_2$(Et$_2$PCH$_2$CH$_2$PEt$_2$)$_2$].

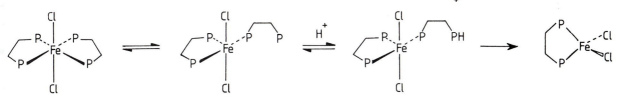

Fig. 5.13

In this reaction the acid is not involved in the dissociation of the chelate ring, rather the initial dissociation occurs without the intervention of protons and is occurring all the time in solution. In the absence of acid the pendant donor atom of the ring-opened chelate rapidly returns to metal. However, in the presence of acid, protonation of the pendant donor atom can effectively compete with the ring-closure reaction, and by binding to the lone pair of electrons, blocks the binding to the metal, leading ultimately in the loss of the chelate. This type of reaction is very general and is found for a wide range of metals, oxidation states and coordination geometries.

Another general class of acid catalysed reaction is typified by the reactions of *trans-*[M(O)(OH)(CN)$_4$]$^{2-}$ (M = Mo, W, Tc or Re) shown in Fig. 5.14.

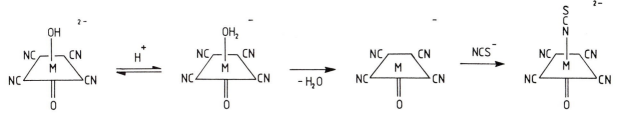

Fig. 5.14

Here, protonation of the coordinated hydroxide generates the labile aquo-ligand. Rate-limiting dissociation of the aquo-ligand is followed by rapid capture of the five-coordinate intermediate by the thiocyanate ion.

Protonation of a non-labile ligand

The electronic influence of the proton can be felt at some distance from its binding site, and in this way protonation of one ligand can lead to labilization of another. This behaviour is seen in the protonation of a

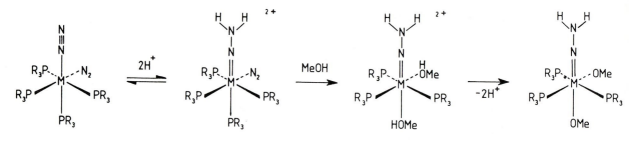

Fig. 5.15

dinitrogen ligand in *cis*-[M(N$_2$)$_2$(PMe$_2$Ph)$_4$] (M = Mo or W), as shown in Fig. 5.15. Here diprotonation of one dinitrogen ligand (to form a hydrazide ligand) has the effect of withdrawing electron-density from the metal centre and thus labilizing the other dinitrogen ligand and a tertiary phosphine ligand, both of which bond to the metal by π-backbonding from the metal d orbitals to vacant orbitals on the ligand. The decrease in electron-density at the metal upon protonation is reflected in an increased lability of these residues. Indeed the whole basis of this type of labilization is such that it will only operate on ligands that bind to metals by use of synergic bonding.

Upon dissociation of the dinitrogen and phosphine ligands, the solvent, methanol, can bind to the metal. In so doing, the two coordinated methanol residues each release a proton, thus regenerating the two protons consumed in the initial activating step, and fulfilling the catalysis. Clearly the use of a protic solvent is essential in systems such as this if the reaction is to be catalytic in protons; in an aprotic solvent similar activation of coordinated dinitrogen can occur but the reaction will consume protons. One final point about the mechanism shown in Fig. 5.15 is worthy of comment. The dissociation of dinitrogen and phosphine and their replacement by the more electron-releasing methoxy-residues changes the acidity of the hydrazido(2-)-residue from a relatively strong acid to a relatively weak acid; the same effect as we saw in the base catalysed substitution of [Mo(NH)X(Ph$_2$PCH$_2$CH$_2$PPh$_2$)$_2$]$^+$.

Protonation at the metal

There are, surprisingly, few examples where protonation of the metal centre labilizes the system to substitution. One example is the acid catalysed exchange of the carbonyl ligands in [Fe(CO)$_5$], where it has been proposed that formation of [FeH(CO)$_5$]$^+$ labilizes the carbonyl ligands to dissociation. This interpretation is however not unambiguous. In the reactions of acid with complexes containing metals in very low oxidation states, such as in this example, due consideration must be given to the possibility that the acid is behaving not as a proton source but as an oxidant. The effect of one electron oxidizing [Fe(CO)$_5$] to [Fe(CO)$_5$]$^+$ on the lability of the carbonyl ligands would be similar to that of protonation.

We will meet other examples of protonation at a metal in Chapter 8.

5.3 Redox catalysed substitution

Changing the oxidation state of a metal can alter the lability of that metal dramatically. It is by making use of this redox property of metal complexes that the substitution reaction can be accelerated. This is best explained by the two examples described below.

Chromium(III) complexes are substitutionally robust, but the addition of catalytic amounts of chromium(II) results in the rapid substitution of the chromium(III) species by the mechanism shown in Fig. 5.16.

The chromium(II) species is substitutionally labile and hence binds X^- rapidly. Subsequent electron transfer to the $Cr^{III}Y$ complex generates the $Cr^{III}X$ species (which is now substitutionally robust) and the labile $Cr^{II}Y$. The overall effect is the rapid substitution of Y for X at the chromium(III) site.

$$
\begin{array}{lll}
\overset{III}{Cr}\!-\!Y \;+\; X \;\longrightarrow\; \overset{III}{Cr}\!-\!X \;+\; Y & & slow \\[4pt]
\overset{II}{Cr} \;+\; X \;\rightleftharpoons\; \overset{II}{Cr}\!-\!X & & rapid \\[4pt]
\overset{II}{Cr}\!-\!X \;+\; \overset{III}{Cr}\!-\!Y \;\longrightarrow\; \overset{III}{Cr}\!-\!X \;+\; \overset{II}{Cr}\!-\!Y & & rapid \\[4pt]
\overset{II}{Cr}\!-\!Y \;\rightleftharpoons\; \overset{II}{Cr} \;+\; Y & & rapid
\end{array}
$$

Fig. 5.16

In a similar vein, the substitution at octahedral platinum(IV) sites is catalysed by square planar platinum(II) species by the mechanism shown in Fig. 5.17.

$$
\begin{array}{l}
\overset{II}{Pt} \;+\; Y \;\rightleftharpoons\; \overset{II}{Pt}\!-\!Y \\[4pt]
\overset{II}{Pt}\!-\!Y \;+\; \overset{IV}{Pt}\!-\!Cl \;\rightleftharpoons\; Y\!-\!\overset{II}{Pt}\!-\!Cl\!-\!\overset{IV}{Pt} \;\rightleftharpoons\; Y\!-\!\overset{IV}{Pt}\!-\!Cl \;+\; \overset{II}{Pt}
\end{array}
$$

Fig. 5.17

PART TWO: ELECTRON TRANSFER REACTIONS

6 Electron transfer mechanisms

Electron transfer reactions are a dominant feature of transition metal chemistry, and are of fundamental importance in defining the characteristics of this area of the periodic table. The mechanisms of electron transfer between two metal complexes have received considerable attention and two distinct types of mechanisms operate: the inner sphere and the outer sphere pathways. The distinction between the two mechanisms resides in whether the reactants share a ligand during the act of electron transfer (as they do in the inner sphere pathway) or not (as in the outer sphere route). Clearly, in the inner sphere electron transfer mechanisms, substitution processes are an important component of the overall reaction. We are now familiar with substitution mechanisms and hence we will start the discussion of the mechanisms of electron transfer by relating the inner sphere pathway, and then move on to the mechanism of the outer sphere pathway. This order of presentation has several advantages. First, we start with a mechanism which contains elements with which we are familiar, and secondly the detailed theoretical treatments of electron transfer reactions, which relate more to the outer sphere mechanism can be delayed until we are familiar with all the nuances of this reaction type.

6.1 Barriers to electron transfer between metal sites

The majority of the reactions with which we will be concerned are those between two complexes, in which the reductant transfers a single electron to the oxidant, as illustrated by the general equation in Fig. 6.1.

However it is important to be clear from the outset that there are certain restrictions on the transfer of this electron.

The most obvious barrier to electron transfer is the distance that the electron has to travel. The intervening solvent molecules and ligands act as insulation to the two participants and thus the reactants must approach one another sufficiently closely to transfer an electron. Irrespective of the

$$M_{ox} + M'_{red} \longrightarrow M_{red}+ \ M'_{ox}$$

Fig. 6.1

intimate details of the mechanism of electron transfer the close contact of the redox partners minimizes the barrier to electron transfer arising from the solvent and maximizes the electronic coupling between the oxidant and reductant. However even if the two species can approach one another closely enough to transfer an electron, and the reaction is thermodynamically favourable (as determined from the relative redox potentials, E^o as defined in Fig. 6.2), there still remain significant problems.

$$M_{red} \rightleftharpoons M_{ox} + e, \quad E^o$$

Fig. 6.2

Probably the most important barrier to electron transfer has its origins in the bond distances of the oxidant and reductant. Consider the thermodynamically neutral reaction between $[Fe(OH_2)_6]^{3+}$ and $[Fe(OH_2)_6]^{2+}$, as shown in Fig. 6.3.

$$\overset{*}{[Fe(OH_2)_6]}^{2+} + [Fe(OH_2)_6]^{3+} \rightleftharpoons \overset{*}{[Fe(OH_2)_6]}^{3+} + [Fe(OH_2)_6]^{2+}$$

Fig. 6.3

This type of reaction is known as a self-exchange reaction since there is no nett chemical change. In order to follow reactions such as this experimentally, it is necessary to be able to distinguish between the two partners without effecting the process in any way, this is accomplished by isotopically labelling one of the reagents at an iron atom.

The larger effective nuclear charge of the Fe(III) complex means that the equilibrium bond length to the water ligands (Fe^{III}–O = 2.05 A) is significantly shorter than that for the Fe(II) species (Fe^{II}–O = 2.21 A). Since the Franck–Condon restrictions dictate that atom movements are much slower than the transfer of electrons, if electron transfer occurred with the complexes in these equilibrium configurations then it would give rise to products with the bond distances associated with the other oxidation state. This situation cannot prevail and hence both redox partners must change their bond lengths to an intermediate distance prior to the transfer of the electron. After the redox reaction, the two partners can readjust their bond distances to that corresponding to their respective oxidation states. During these reorganizations of the coordination spheres, the solvent molecules in the solvation shell are not inactive, but also have to reorganize themselves to reflect the changing bond lengths during the reaction profile and the change of charge on the complex when the reaction is accomplished.

Another restriction on electron transfer is that there can be no change in the spin state of the electron when it is transferred. Most of the time there is no major problem in this respect. For instance in the self-exchange reactions of $[Co(phen)_3]^{h+}$, the Co(III) and the Co(II) species have a low spin electron configuration, and the electron is transferred from an e_g orbital in the reductant to an e_g orbital in the oxidant. However in most cobalt(II) complexes the electron configuration is high spin, whereas the cobalt(III) complexes are low spin.

As shown in Fig. 6.4 this can lead to problems upon electron transfer. If an electron in the e_g orbital of the Co(II) species is transferred to an e_g orbital in the Co(III) complex, then we end up with strange electron configurations for both products: a Co(II) species with $t_{2g}^6 e_g^1$ (the low spin configuration) and a Co(III) species with $t_{2g}^5 e_g^1$, a configuration in which incompletely filled t_{2g} and e_g orbitals are present!

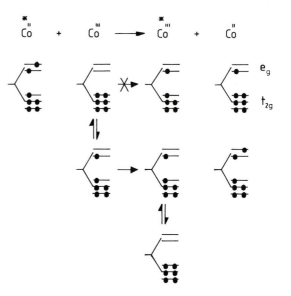

Fig. 6.4

In order to overcome this problem Co(III) complexes must change their electron configuration prior to electron transfer from $t_{2g}^6 e_g^0$ to $t_{2g}^5 e_g^1$. Once in this excited state electron transfer can occur to give a Co(II) species with the correct electron configuration. The necessity that electron transfer to Co(III) complexes only occurs when the bond lengths have reached an intermediate distance between that of the oxidized and reduced forms, together with the change in spin states of the oxidant, leads to low rates of electron transfer for these complexes.

6.2   The inner sphere mechanism

The inner sphere mechanism of electron transfer is shown in Fig. 6.5.

In this mechanism the initial diffusion-controlled association of the reductant and oxidant is followed by a substitution reaction in which a ligand on one of the reactants (usually the oxidant) penetrates the coordination sphere of the other to form the precursor complex. Thus it is

$$M_R^{'} \ + \ M_O^{-}X \ \rightleftharpoons \ M_R^{'}.M_O^{-}X$$

$$M_R^{'}.M_O^{-}X \ \rightleftharpoons \ M_R^{'}-X-M_O$$

$$M_R^{'}-X-M_O \longrightarrow \ M_O^{'}-X-M_R$$

$$M_O^{'}-X-M_R \longrightarrow \ M_O^{'}-X \ + \ M_R \ \underline{or} \ M_O^{'} \ + \ M_R^{-}X \ \underline{or} \ M_O^{'} \ + \ M_R \ + \ X$$

| Precursor complex |
| Postcursor complex |

Fig. 6.5

essential that at least one of the reactants is either coordinatively unsaturated, such as $[Co(CN)_5]^{3-}$, or at least labile to dissociation, such as $[Cr(OH_2)_6]^{2+}$.

Up to the formation of the precursor, the reaction has only the characteristics of the substitution reactions we saw in the earlier section, in which a ligand on one partner acts as a nucleophile for the other metal centre, and no transfer of electrons has occurred yet. The only prerequisite for the bridging ligand is that it must have a means of binding to the new metal site. Once the precursor complex has been formed then electron transfer can occur from the reductant to the oxidant, thus generating the postcursor complex, which is identical to the precursor species except for the relative oxidation states of the two metal sites. Finally the postcursor complex can now dissociate to give the products.

The first demonstration of the inner sphere mechanism was in the reaction between $[Cr(OH_2)_6]^{2+}$ and $[CoCl(NH_3)_5]^{2+}$, as outlined in Fig. 6.6.

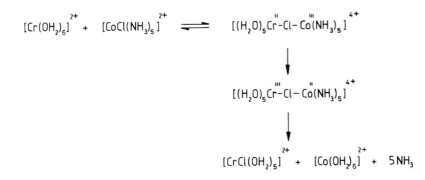

Fig. 6.6

The success of this demonstration resides in the relative substitution labilities of the reactants and the products. The Cr(II) reductant is extremely labile, whereas the Co(III) species is substitutionally robust. Upon transfer of the electron the resulting Cr(III) product is substitutionally robust and the Co(II) site is labile. This has the consequence, as shown in Fig. 6.6, that the chloro-group which is brought into the precursor complex by the Co(III) species, leaves the postcursor complex with the Cr(III) site. In order to prove the intervention of the inner sphere pathway, it was essential not only to show that the $[CrCl(OH_2)_6]^{2+}$ is formed quantitatively, but also to show that $[Cr(OH_2)_6]^{3+}$ incorporates Cl^- from solution much more slowly than the production of $[CrCl(OH_2)_5]^{2+}$ in the redox reaction, thus ensuring that the chloride is not incorporated into a Cr(III) species after the redox reaction. In addition it must be shown that the solvolysis of the Co(III) complex is slow compared to its reduction.

Other examples in which ligand transfer from the oxidant to the reductant has been observed include the reaction of $[FeCl(OH_2)_5]^{2+}$ with $[Cr(OH_2)_6]^{2+}$ and the reaction of $[V(OH_2)_6]^{2+}$ with $[Co(N_3)_2en_2]^+$ (en = $NH_2CH_2CH_2NH_2$). In this last example the oxidized product, $[V(N_3)(OH_2)_5]^{2+}$, is detected as the short-lived product of the electron transfer reaction before it solvates to $[V(OH_2)_6]^{3+}$.

If we consider the generalized inner sphere mechanism shown in Fig. 6.5, it is clear that the rate-limiting step can be any of the following.
(1) The formation of the precursor complex;
(2) the transfer of the electron;
(3) the dissociation of the postcursor complex.

Assigning which step is rate-limiting is not usually possible from the form of the rate law in electron transfer reactions. Nearly always, the rate of these reactions exhibits a simple first order dependence on the concentration of both the reductant and the oxidant. Consequently other criteria must be employed to establish which is the slowest step in the mechanism.

Formation of the precursor complex

Electron transfer reactions which are rate-limited by the formation of the precursor complex are those involving species whose substitution lability is very low, as typified by the reactions of $[V(OH_2)_6]^{2+}$. Notice here that V^{2+} is a d^3 ion and is consequently expected to be kinetically robust (see Chapter 4). In this class of reactions, the rate of the overall redox reaction is independent of the thermodynamic driving force of the redox reaction (measured by the difference in the redox potentials for the two half-reactions of the reacting couple, $\Delta E = E^o_{ox} - E^o_{red}$), and, for a dissociatively activated process, the nature of the bridging ligand. In addition, of course, the rate of the reaction is close to that of the rate of solvent exchange.

In practice, the substitution rate at the metal site may be so slow that the complex opts to transfer an electron by an outer sphere pathway. In such a situation it becomes a problem in deciding (for a series of reactions) which are true inner sphere reactions.

Rate-limiting electron transfer

The electron transfer step will be rate-limiting in systems in which the formation of the precursor complex and decomposition of the postcursor complex are both rapid; that is when at least one of the metal sites is substitutionally labile. This is typified by the large number of reactions studied with various cobalt(III) complexes and $[Cr(OH_2)_6]^{2+}$, such as that in Fig. 6.6.

Under advantageous conditions, the precursor complex can be detected and the subsequent slow electron transfer monitored. This has been observed in the reaction between $[Fe(CN)_5(OH_2)]^{3-}$ and the cobalt(III) complex shown in Fig. 6.7.

$$[Fe(CN)_5OH_2]^{3-} + [Co(NH_3)_5(N \bigcirc N)]^{3+} \longrightarrow [(NC)_5Fe-N\bigcirc N-Co(NH_3)_5]$$

precursor complex

Fig. 6.7

We are rarely able to detect precursor complexes, but it is possible to investigate electron transfer within this type of binuclear species by alternative means. Binuclear complexes such as those shown in Fig. 6.8 can be prepared independently in which both metal sites are oxidized. Subsequent reaction of this species with a reducing agent such as $[Ru(NH_3)_6]^{2+}$ or Eu(II) selectively reduces the Ru(III) site in the binuclear complex, and the irreversible transfer of the electron to the Co(III) site can then be monitored.

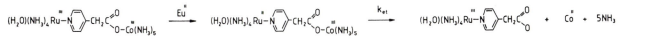

Fig. 6.8

Studies on systems such as these give detailed information about the transfer of electrons within inner sphere complexes. In particular, by varying the separation of the metal centres, it is found, as expected, that the free energy of activation for electron transfer ($\Delta G^{\ddagger}$) varies inversely with the distance between the two metals.

Role of the bridge in electron transfer

A question of fundamental importance in understanding the details of the inner sphere pathway for electron transfer is the role played by the bridging ligand. Is the bridge the mediator of the electron as it travels from one metal to the other or is it just a means of keeping the two sites close to one another? Certainly there are some systems such as that between $[Cr(OH_2)_6]^{2+}$ and *cis*-$[Co(N_3)_2en_2]^+$ where a double bridge is formed between the reductant and the oxidant, and clearly not both bridges are required to mediate the transfer of the single electron.

The two possible intimate mechanisms of electron transfer within the precursor complex are as follows.

Resonance transfer: in which the electron is transferred directly from one metal to the other by a tunnelling mechanism.

Chemical transfer: in which the electron is transferred initially to the bridge and then onto the oxidant, as shown in Fig. 6.9.

$$M'_R-X-M_O \longrightarrow M'_O-\dot{X}-M_O \longrightarrow M'_O-X-M_R$$

Fig. 6.9

Various lines of evidence now favour the chemical transfer mechanism in systems containing reducible ligands. This evidence includes the following.

1. The detection of radical cations by EPR spectroscopy during the electron transfer reaction, in which the electron can be shown to be on the bridging ligand.

2. The rate of electron transfer to complexes containing Co(III)–L and the conjugate acid of L (H–L$^+$) occur at much the same rate.

A distinction between the chemical transfer and resonance transfer mechanisms is possible using isotope effects. By comparing the rates of electron transfer to $[CoL(NH_3)_5]^{n+}$ and $[CoL(ND_3)_5]^{n+}$ (when L is non-reducible, L = HO, CN, NCS or MeCO$_2$) the isotope ratio, k_H / k_D, is in the range 1.3 to 1.5. This is consistent with the reorganization of the coordination sphere prior to electron transfer (Franck–Condon restrictions). However for the complexes in Fig. 6.10 containing reducible ligands the isotope ratio is much smaller, $k_H / k_D = 1.1 \pm 0.02$, consistent in these cases with rate-limiting transfer to the reducible ligand.

The advantages of the chemical transfer mechanism over the resonance transfer pathway are two-fold. First, the Franck–Condon restrictions do not have to be met by both metal centres at the same time during chemical transfer, thus leading to an increased rate of electron transfer. Secondly, a direct interaction between the oxidant and the reductant means that an electron transferring between them has to overcome the activation barriers for both the release of the electron from the reductant and the acceptance of the electron by the oxidant, all in one action. In contrast, for the chemical transfer pathway, these same barriers are met successively, and hence the energy required at each stage of the reaction profile is correspondingly less.

$[Co(NH_3)_5L]^{3+}$

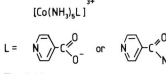

Fig. 6.10

Remote attack in electron transfer

For an inner sphere mechanism, the two metal sites can be separated by several atoms and still transfer electrons at an appreciable rate, provided the ligand system that bridges the oxidizing and reducing centres is unsaturated, and hence conducting. This concept of electron transfer at a distance mediated by the bridge is enforced on the system, such as in the reduction of $[Ru(NH_3)_5(\text{iso-nicotinamide})]^{3+}$ shown in Fig. 6.11, since it is only the remote atoms which contain a lone pair of electrons for binding the reductant.

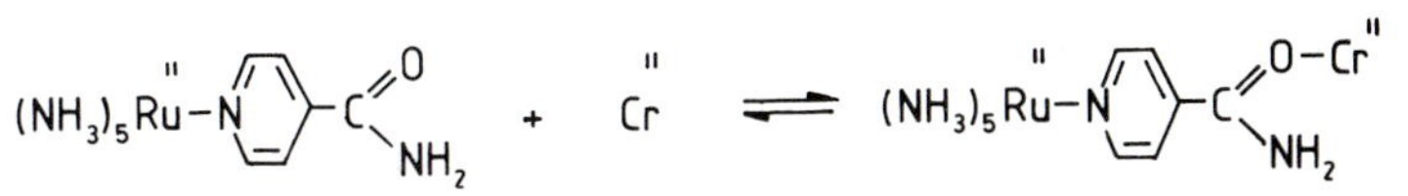

Fig. 6.11

Remote attack has also been demonstrated in the reaction between $[Fe(\underline{N}CS)(OH_2)_5]^{2+}$ and $[Cr(OH_2)_6]^{2+}$ where the product is $[Cr(\underline{S}CN)(OH_2)_5]^{2+}$. However it should not be assumed that the metals will always want to be at extreme ends of an unsaturated molecule and rely on the ligand to mediate the transfer of the electron. For instance in the reaction of $[Co(N_3)(NH_3)_5]^{2+}$ with $[Cr(OH_2)_6]^{2+}$ the reductant binds not at the azido-nitrogen atom remote from the cobalt, but at the nitrogen atom bound to the cobalt.

Rate-limiting postcursor dissociation

Having brought the reactants together and transferred the electron, the only stage left to be rate-limiting is the dissociation of the postcursor complex. This stage will be rate-limiting if both the metals, in their new oxidation states, are substitutionally robust. This happens in the reaction shown in Fig. 6.12 where the postcursor complex is isolable.

Fig. 6.12

In a similar vein, a postcursor complex can be detected in the reaction of $[Ru(NH_3)_5(iso\text{-}nicotinamide)]^{3+}$ with $[Cr(OH_2)_6]^{2+}$. The identification of this detected species as the postcursor, rather than the precursor complex, relies on its spectroscopic and reactivity properties. Thus the complex shows strong π^*–d absorptions in the electronic spectrum consistent with a Ru(II) coordinated to the pyridine residue, but inconsistent with a Ru(III) species. In addition the intermediate exhibits the characteristic d–d absorption spectrum of a Cr(III) entity in an oxygen-ligated environment, and the aquation characteristics of a Cr(III) species. With this evidence then, despite the fact that the reaction involves no transfer of the bridging ligand as a consequence of electron transfer, the nature of the reaction can still be identified unambiguously.

Dead-end species: a warning about detected intermediates

At first sight, the reaction between $[Fe(CN)_6]^{3-}$ and $[Co(edta)]^{2-}$ (edta $=$ $(O_2CCH_2)_2NCH_2CH_2N(CH_2CO_2)_2^{4-}$) looks as if it falls in the category discussed above. Study of this reaction reveals an accumulation of the binuclear intermediate shown in Fig. 6.13, apparently a postcursor complex.

$$[Fe(CN)_6]^{3-} + [Co(edta)]^{2-} \rightleftharpoons [(NC)_5Fe\text{-}CN\text{-}Co(edta)]^{5-}$$

$$\downarrow outer\ sphere$$

$$[Fe(CN)_6]^{2-} + [Co(edta)]^{-}$$

Fig. 6.13

However detailed kinetic arguments demonstrate that this species is not on the pathway leading to products. Rather it is a dead-end species which accumulates in significant concentrations but does not react further to give the redox products. The mechanism by which an electron is transferred in this system is the outer sphere pathway. Any material present as the binuclear dead-end species must dissociate to reactants and then proceed through the outer sphere mechanism to give products.

The important message for all mechanistic studies is that just because a species accumulates during the reaction, does not ensure that this species is an intermediate on the pathway to products. Even the form of the rate law in the mechanism shown in Fig. 6.13 suggests that the binuclear is an intermediate on the pathway, when in fact, the situation is different.

6.3 The outer sphere mechanism

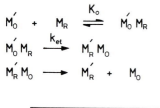

$$M_O' + M_R \overset{K_o}{\rightleftharpoons} M_O' M_R$$

$$M_O' M_R \overset{k_{et}}{\longrightarrow} M_R' M_O$$

$$M_R' M_O \longrightarrow M_R' + M_O$$

$$\boxed{k_{obs} = \frac{k_{et} K_o [M_O']}{1 + K_o [M_O']}}$$

Fig. 6.14

When at least one of the complexes undergoing electron transfer is substitutionally robust and does not contain a ligand capable of acting as a bridge to the other reactant, the electron transfer is enforced to proceed via the outer sphere pathway shown in Fig. 6.14.

In this mechanism the rapid diffusion of the two reagents together forms the outer sphere adduct which then undergoes electron transfer to yield the products.

The outer sphere mechanism is ensured in the reaction between two coordinatively saturated complexes such as in the reaction between $[Ru(NH_3)_6]^{2+}$ and $[Ru(NH_3)_5(pyridine)]^{3+}$, or the oxidation by Cr(III) complexes or $[Co(o\text{-phenanthroline})_3]^{3+}$ of a variety of metal carbonyl anions, $[Re(CO)_5]^-$, $[Co(CO)_4]^-$, $[Mo(\eta^5\text{-}C_5H_5)(CO)_3]^-$, $[Fe(\eta^5C_5H_5)(CO)_2]^-$ and $[Mn(CO)_4L]^-$ (L = CO or PR_3). This last series of studies demonstrates the insensitivity of the reaction mechanism to the coordination number of the reactants.

The outer sphere mechanism is also ensured in the reactions of species such as $[Cr(OH_2)_6]^{2+}$ with $[Coen_3]^{3+}$ (en = $NH_2CH_2CH_2NH_2$). Although the Cr(II) centre is substitutionally labile, the robust nature of the Co(III) site together with the lack of any potential bridging ligand, precludes an inner sphere complex being formed. In addition, as we saw in the reactions of $[V(OH_2)_6]^{2+}$, when one of the redox partners contained a potential bridging ligand but the other species is only borderline labile, the mechanistically simpler outer sphere pathway may be the dominant route. In the absence of any complicating features the outer sphere pathway would be expected to be faster than the inner sphere pathway, since the energetic demands of the former mechanism are less.

The limiting rate law

The most commonly observed rate law for outer sphere electron transfer reactions between two metal sites is a simple first order dependence on the concentration of both the oxidant and the reductant. This rate law is the limiting form of the generalized expression shown in Fig. 6.14, under conditions where there is no significant accumulation of the outer sphere adduct, $1 \gg K_o[M_o']$, when the reaction is studied in the presence of a large excess of the oxidant. However, the general form of the rate law shown in Fig. 6.14 is observed for redox reactions between species of large and opposite charge, since the large electrostatic interaction between the two species leads to an accumulation of the outer sphere adduct.

The general form of the rate law is observed in reactions of the type shown in Fig. 6.15 between oxidants with a 2+ or 3+ charge and the $[Fe(CN)_6]^{4-}$ reductant.

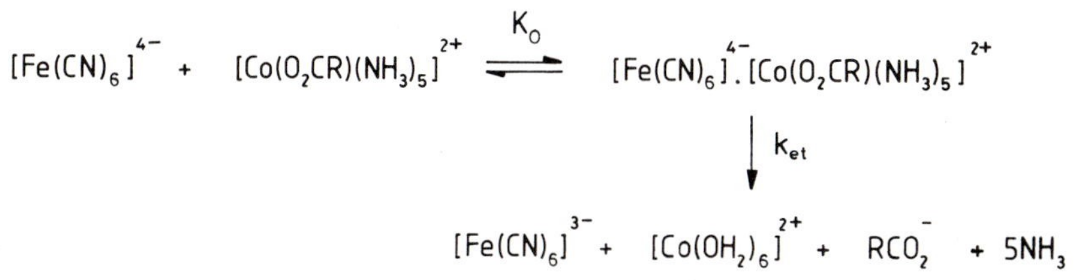

$$[Fe(CN)_6]^{4-} + [Co(O_2CR)(NH_3)_5]^{2+} \underset{}{\overset{K_o}{\rightleftharpoons}} [Fe(CN)_6]^{4-}.[Co(O_2CR)(NH_3)_5]^{2+}$$

$$\downarrow k_{et}$$

$$[Fe(CN)_6]^{3-} + [Co(OH_2)_6]^{2+} + RCO_2^- + 5NH_3$$

Fig. 6.15

With the values of the independent equilibrium and rate constants available it is possible to probe the characteristics of the adduct formation and the electron transfer step in isolation. Thus the association constants for the formation of the outer sphere adduct are in good agreement with those values derived from calculations based on simple electrostatic arguments. The rate of the electron transfer step within the outer sphere complex depends upon the nature of the substituent, R, on the carboxylate group. The rate of electron transfer increases as the pK_a associated with these residues decreases. The decreasing pK_a through this series is associated with these residues becoming progressively better electron-withdrawing ligands. The same general trend, in the rates of electron transfer with the nature of the substituents, has been observed in the reactions of these same oxidants with the reductant, $[Ru(NH_3)_6]^{2+}$. In this case the formation of the outer sphere adduct is not observed but the value of K_o can be estimated from simple electrostatic arguments.

Geometry of the outer sphere adduct

Although the stereochemical requirements of the outer sphere mechanism are less stringent than those of the inner sphere pathway, it would be wrong to consider that the outer sphere adduct is entirely devoid of stereochemical interactions between the two reactants. These reactants are not like two hard spheres; they have a geometry which can be highly symmetrical (e.g. $[Co(NH_3)_6]^{3+}$) or one with distinctly different areas associated with different ligands (e.g. $[CoX(NH_3)_5]^{n+}$, where X can be a neutral or anionic ligand, the ligand itself consisting of a single atom or a great many atoms).

The forces which hold the outer sphere adduct together will be a combination of electrostatic interactions, van der Waal's forces and hydrogen bonding.

Consideration of simple theoretical calculations and X-ray crystal structures such as that of Δ-$[Co\{lel$-(-)-$pn\}_3]^{3+}$. Δ-$[Cr(mal)_3]^{3-}$ (pn = $H_2NCH_2CH_2CH_2NH_2$, mal = malonate) indicates that although the outer sphere adduct (ion pair) has a dynamic geometry, there are dominant hydrogen-bonding interactions. These specific interactions are illustrated in the examples in Fig. 6.16, and involve hydrogen bonds along mutual C_3 axes, C_3C_2 axes and C_2C_2 axes (not illustrated).

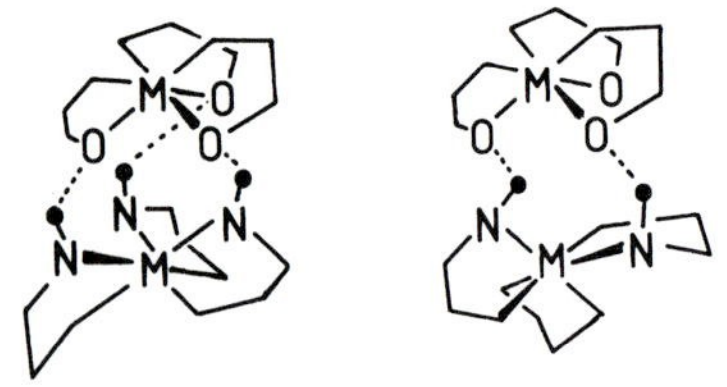

Fig. 6.16

6.4   Theoretical treatment of electron transfer

Reactions which proceed by the outer sphere mechanism are well suited to theoretical treatment since there is no bond-making or bond-breaking processes involved. The treatment of these reactions, developed predominently by Marcus and Hush, involve calculating the free energy of change (ΔG^*) to bring the reactants together, from infinite separation, to form the activated, outer sphere complex. That is not just the energy to bring the two species together, but also to perturb their coordination and solvation spheres in line with the Franck–Condon restrictions. Also, the treatment assumes that the transfer of the electron is adiabatic, that is, the transfer of the electron does not involve the loss of energy from the system.

The value of ΔG^* is made up of three separate components:

(1) the work necessary to bring the reactants together, and then separate them after the reaction (w^r and w^p);

(2) the energies to reorganize the coordination spheres and the solvation spheres of the two reactants (λ_i and λ_o respectively);

(3) the standard free energy change of the reaction (that is the thermodynamic driving force of the reaction, ΔG°).

The most difficult of these terms to estimate is the values for the reorganization of the ligands, λ_i, prior to electron transfer.

We are not going to get bogged down in all the intricacies associated with this treatment, but just sufficiently to appreciate the more important derived relationships. The most important equations are shown in Fig. 6.17, where h = Planck's constant, k is the Boltzmann constant and Z is the diffusion-controlled collision frequency.

$$\Delta G^* = \frac{w^r + w^p}{2} + \frac{\lambda_o + \lambda_i}{4} + \frac{\Delta G^\circ}{2} + \frac{(\Delta G^\circ + w^p - w^r)}{4(\lambda_o + \lambda_i)}$$

$$\Delta G^* = \Delta G^{\ddagger} - RT\ln(hZ/kT)$$

$$k = \kappa Ar^2 \exp(-\Delta G^*/RT)$$

Fig. 6.17

Of particular importance are the relationships between ΔG^* and $\Delta G^{\ddagger}$ and the rate constant for electron transfer, $\underline{k}$. In the equation relating k and ΔG^* the transmission coefficient, κ is approximately 1 for adiabatic reactions, and the term Ar^2 is the effective collision frequency and approximates to $Z = 10^{11} \, dm^3 mol^{-1} s^{-1}$. Using these relationships, reasonable fits to the outer sphere electron transfer self-exchange rates are obtained. For example, the self-exchange rate constant for the $[Tc(Me_2PCH_2CH_2PMe_2)_3]^{+/2+}$ couple is found experimentally to be $k_{exp} = 0.6 \times 10^6 \, dm^3 mol^{-1} s^{-1}$ and calculated to be $k_{calc} = 3 \times 10^6 \, dm^3 mol^{-1} s^{-1}$. Considering the approximations necessary to arrive at the equations in Fig. 6.17 this agreement in order of magnitude is very good.

It is important that self-exchange rate constants are determined directly to complement the data from the theoretical treatment. These rate constants can be determined using line-broadening in ^1H, ^{13}C, etc. NMR spectra. In the example shown in Fig. 6.18, ^{59}Co NMR. spectroscopy was used.

$$[\overset{*}{Co}(S_3)_2]^{3+} + [Co(S_3)_2]^{2+} \longrightarrow [\overset{*}{Co}(S_3)_2]^{2+} + [Co(S_3)_2]^{3+}$$

$$S_3 =$$

Fig. 6.18

Because of the sulfurous ligands, both oxidation states of the cobalt complex are low spin species. Addition of the paramagnetic Co(II) complex leads to broadening of the signals attributable to the diamagnetic Co(III) species. Analysis of the data gives $k_{ex} = 1.3 \times 10^5 \, dm^3 mol^{-1} s^{-1}$, in reasonable agreement with the theoretically derived value of $1 \times 10^6 \, dm^3 mol^{-1} s^{-1}$.

$$k_{12} = \sqrt{k_{11} k_{22} K_{12} f_{12}}$$

$$\log f_{12} = \frac{\log K_{12}}{4 \log(k_{11} k_{22}/Z^2)}$$

Fig. 6.19

Probably the most important approximation of the theoretical equations is given in Fig. 6.19, where $f_{12} = Z^2_{12} / Z_{11} \cdot Z_{22}$, the ratio of the collision frequencies associated with each species in the reaction system.

Using equations we can calculate the rate constant for a thermodynamically-driven outer sphere electron transfer reaction, k_{12}, from a knowledge of the self-exchange rate constants of each reactant and the equilibrium constant K_{12} of the reaction. In particular if $K_{12} \sim 1$, then

$\log f_{12} \sim 0$ is a useful approximation. The value of K_{12} is determined from the redox potentials of the reactants. Agreement between k_{12}^{calc} and k_{12}^{exp} can be used to indicate that a reaction operates by an outer-sphere pathway. Also for a series of reactions of, for instance, a given oxidant with a variety of reductants the graph of $\log k_{12}$ against $\log K_{12}$ will be a straight line if all the reactions operate by the outer sphere mechanism; deviations from the line may be indicative of a change in mechanism.

6.5   Non-complementary electron transfer

All of the discussion so far has concentrated on complementary electron transfer reactions, that is reactions in which the reductant and the oxidant want to give up and accept, respectively, the same number of electrons. Invariably this has been the transfer of a single electron since the energetic requirements to satisfy the Franck–Condon restrictions are too demanding for multi-electron transfer.

The redox chemistry of the metals in the p-block is dominated by stable oxidation states differing by two units (e.g. PCl_3 and PCl_5). Hence the oxidation of p-block complexes by transition metals capable of transferring only a single electron has to proceed through two steps. Each step involves the transfer of a single electron, and the formation of a transient, unstable one electron oxidized p-block element. An example is shown in Fig. 6.20, and analogous reactions of Fe(III) with Tl(I) and Sn(II) are known. In all cases the reaction is rate-limited by the first step, producing the unstable oxidation state of the p-block element: that is the formation of As(IV), Tl(II) and Sn(III), respectively.

Similarly, in the reduction of p-block complexes by transition metal complexes a two-step pathway operates with the slow step again being the production of the unstable oxidation state of the p-block species. This is typified by the example shown in Fig. 6.21.

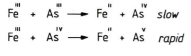

Fig. 6.20

$$[\text{Ru}(\text{NH}_3)_5(\text{bipy})]^{2+} + \text{S}_2\text{O}_8^{2-} \rightleftharpoons [\text{Ru}(\text{NH}_3)_5(\text{bipy})]^{2+}.\text{S}_2\text{O}_8^{2-}$$

$$[\text{Ru}(\text{NH}_3)_5(\text{bipy})]^{2+}.\text{S}_2\text{O}_8^{2-} \longrightarrow [\text{Ru}(\text{NH}_3)_5(\text{bipy})]^{3+} + \text{SO}_4^{2-} + \text{SO}_4^{-} \qquad slow$$

$$[\text{Ru}(\text{NH}_3)_5(\text{bipy})]^{2+} + \text{SO}_4^{-} \longrightarrow [\text{Ru}(\text{NH}_3)_5(\text{bipy})]^{3+} + \text{SO}_4^{2-} \qquad rapid$$

Fig. 6.21

Here the disulfate ion and the Ru(II) cation form a detectable outer-sphere adduct prior to electron transfer. The subsequent electron transfer and dioxygen cleavage step are synchronous, to produce SO_4^{2-} and the unstable SO_4^{-}. The latter species then rapidly reacts with another molecule of the Ru(II) complex.

Of course, non-complementary reactions can also be found between d-block species, for instance the three electron oxidation of Cr(III) to Cr(VI) by $[Fe(CN)_6]^{3-}$ is rate-limited by the initial formation of the Cr(IV) species.

In all these cases the intimate mechanism of the electron transfer can only be defined for the first electron transferred, the transfer of the second electron occurs after the rate-limiting step and thus is not reflected in the kinetics.

A non-complementary reaction can have an important effect on the products formed by the inner sphere mechanism. This is illustrated in the reaction shown in Fig. 6.22.

$$L_5\overset{IV}{Pt}-Cl + \overset{II}{Cr} \longrightarrow L_5\overset{III}{Pt} + \overset{III}{Cr}-Cl$$

$$L_5\overset{III}{Pt} + \overset{II}{Cr} \longrightarrow L_4\overset{II}{Pt} + \overset{III}{Cr} + L$$

Fig. 6.22

The two electron reduction of the Pt(IV) species by Cr(II) is rate-limited by the initial inner sphere pathway to produce the Pt(III) species. But only half of the Cr(III) product is present as $[CrCl(OH_2)_5]^{2+}$, since the subsequent reduction of the Pt(III) species does not involve a chloro-complex.

6.6 Atom transfer pathways

We have seen in the above discussion that a reaction operating by an inner sphere mechanism can involve the transfer of an atom from one metal site to the other. The question then arises as to whether these reactions should really be considered as atom transfer reactions rather than electron transfer.

If we consider the reduction of alkyl halides by metal sites then this problem resurfaces. Two distinct mechanisms can operate in these reactions, as shown in Fig. 6.23.

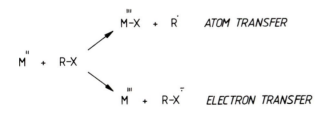

Fig. 6.23

In the atom transfer pathway, the metal abstracts a halogen atom from the alkyl halide and the generated alkyl radical can subsequently react with another molecule of the metal complex. This pathway is analogous to the inner sphere electron transfer mechanism in that it can only be associated with a coordinatively unsaturated or substitutionally labile metal site such as, $[Cr(OH_2)_6]^{2+}$, $[Cren_2]^{2+}$ or $[Co(CN)_5]^{3-}$.

The electron transfer pathway is enforced on substitutionally robust complexes, and involves the transfer of an electron to the lowest unoccupied molecular orbital on the alkyl halide (a C–X σ^* orbital). For instance in the reaction between *trans*-[Mo(CO)$_2$(Me$_2$PCH$_2$CH$_2$PMe$_2$)$_2$] and Ph$_3$CCl the transient Mo(I) intermediate and Ph$_3$C· can be detected by EPR spectroscopy.

Irrespective of the pathway adopted, the kinetics of the reaction between [Cr(OH$_2$)$_6$]$^{2+}$ and alkyl halides generally exhibit a simple first order dependence on the concentrations of both the chromium complex and the alkyl halide. However it is possible to distinguish between atom transfer and electron transfer pathways if the products of the reaction are substitutionally robust. One such example is given in Fig. 6.24.

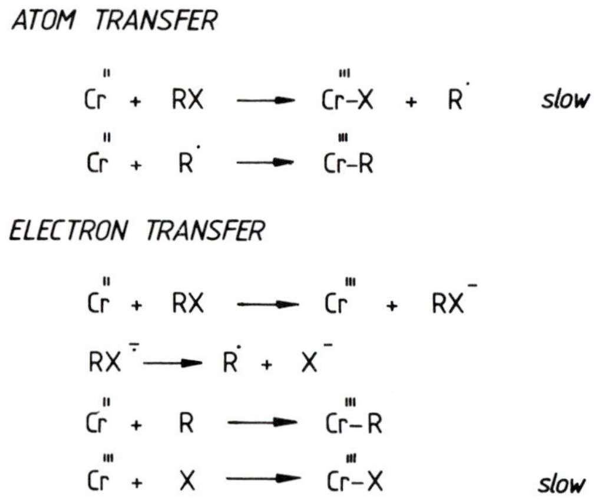

ATOM TRANSFER

$$\overset{II}{Cr} + RX \longrightarrow \overset{III}{Cr}{-}X + R^· \quad slow$$

$$\overset{II}{Cr} + R^· \longrightarrow \overset{III}{Cr}{-}R$$

ELECTRON TRANSFER

$$\overset{II}{Cr} + RX \longrightarrow \overset{III}{Cr} + RX^-$$

$$RX^- \longrightarrow R^· + X^-$$

$$\overset{II}{Cr} + R^· \longrightarrow \overset{III}{Cr}{-}R$$

$$\overset{III}{Cr} + X^- \longrightarrow \overset{III}{Cr}{-}X \quad slow$$

Fig. 6.24

If the mechanism involves atom transfer, then equimolar concentrations of Cr^{III}-X and Cr^{III}-R will be formed. However if the reaction proceeds by electron transfer then equimolar concentrations of Cr^{III} and Cr^{III}-R are obtained, since the initially formed Cr^{III} is substitutionally robust and will only slowly bind with X$^-$.

7 The oxidative-addition reaction

The oxidative-addition reaction is one in which there is a simultaneous increase in the formal oxidation state of the metal and in its coordination number, as shown in a generalized manner in Fig. 7.1.

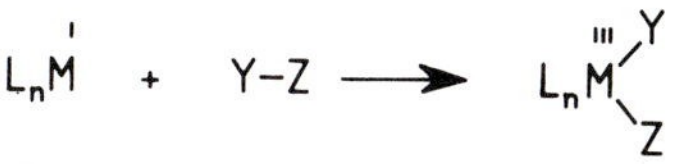

The molecule designated as YZ in this representation can be a wide variety of species, most importantly: H_2, X_2, RSSR, HX, RSH, $SiHR_3$, $SnXR_3$, $HgCl_2$, RCOX, and RX (X = Cl, Br or I). Not all the mechanistic details of all these reactions have been defined. In this chapter we will concentrate on just one system: the oxidative-addition of alkyl halides.

An important aspect of the oxidative-addition reaction is that the term oxidative-addition does not describe the mechanism of a reaction but is merely a stoichiometric definition of the result of the reaction. As we shall see even for the one substrate, RX, the reaction represented in Fig. 7.1 can be accomplished by very different mechanisms.

The types of complexes that undergo the oxidative-addition reaction fall into a relatively narrow group. In general the complexes contain a metal centre with a formal d^8 or d^{10} electron configuration, but in addition are coordinatively unsaturated or potentially so, because of their lability. Thus the reactions are restricted to the following: Ru(0), Os(0), Rh(I), Ir(I), Pd(0 or II), and Pt(0 or II). In addition, whereas complexes such as *trans*-IrCl(CO)(PPh$_3$)$_2$] are square-planar and formally have two preformed vacant sites ready for interacting with the addition molecule, the species Pd(PPh$_3$)$_3$] must dissociate a triphenylphosphine ligand before it can react.

$$L_nM^I \; + \; Y-Z \; \longrightarrow \; L_nM^{III} \underset{Z}{\overset{Y}{\diagup}}$$

Fig. 7.1

7.1 Oxidative-addition by nucleophilic displacement

Conceptually, the simplest interaction between a metal site and an alkyl halide which results in the oxidative-addition of the latter, is one in which the metal site acts as a nucleophile with respect to the carbon centre, as shown on the left-hand of Fig. 7.2. This interaction is analogous to the classic S_N2 reaction in organic chemistry. However, with a metal site acting as the nucleophile other types of interactions are possible, most notably that shown in Fig. 7.2 in which the metal can interact not only with the carbon centre but also with the halide. In the analogous oxidative-addition reactions of H_2 the initial interaction with the metal is almost certainly analogous to that shown in the right-hand structure.

These two different geometries for the transition state of the reaction

$$\begin{array}{c} X \\ | \\ R \\ | \\ M \end{array} \qquad \qquad \begin{array}{c} R-X \\ \diagdown \diagup \\ M \end{array}$$

Fig. 7.2

with RX will have different stereochemical consequences on the reaction, both with respect to the alkyl group and the metal site.

Nucleophilicity of the metal site

A wide variety of complexes can act as a nucleophile towards alkyl halides, most notably the following. In group VI: $[M(\eta^5\text{-}C_5H_5)(CO)_3]^-$ (M = Cr, Mo or W); in group VII: $[M(CO)_5]^-$ (M = Mn or Re); in group VIII: $[M(\eta^5\text{-}C_5H_5)(CO)_2]^-$ (M = Fe or Ru); in group IX: $[M(\eta^5\text{-}C_5H_5)(CO)(PR_3)]$ (M = Co, Rh or Ir) and $[Co(dmgH)_2L]^-$ (dmgH = ONC(Me)C(Me)NOH), and finally in group X: $[M(PR_3)_3]$ (M = Ni, Pd or Pt). Within each group the relative nucleophilicities towards alkyl halides varies down the group, however there is no uniform trend. For instance, in group X the relative nucleophilicities are in the order Ni > Pd > Pt, whereas in group IX the order is the opposite, Ir > Rh > Co.

These metal-based complexes are some of the strongest nucleophiles known, and this has led to them being termed 'supernucleophiles'. It is very difficult to define a scale of nucleophilicity for all these metal species because of the marked influence of the solvent on the nucleophilicity, and the problem of ensuring that the complex is always acting as a nucleophile rather than switching over to a free radical mechanism, which will be discussed later. However it has been possible to determine the nucleophilicity of some of these complexes. Using the Pearson scale of nucleophilicity, $n = k_N / k_O$ where k_N is the second order rate constant for the attack of the nucleophile on MeI and k_O is the corresponding rate constant for the attack of methanol on MeI. The 'supernucleophilicity' of the complex, $[Co(dmgH)_2L]$ is reflected in the value $n = 12$–14; this is to be compared with the value for the strong, conventional nucleophiles, CN^- or I^- for which $n = 7$.

It is impossible to generalize about the relative nucleophilicities of even a single class of metal complex since the ligands have a profound influence on the nucleophilicity. Thus $[RhCl(PPh_3)_3]$ reacts rapidly with MeI, but simple replacement of one of the phosphine ligands by the more electron-withdrawing carbon monoxide renders *trans*-$[RhCl(CO)(PPh_3)_2]$ only poorly nucleophilic towards MeI. Even less drastic changes in the ligand environment of a complex changes the nucleophilicity of the metal centre. This effect has been most extensively studied in $[Co(dmgH)_2L]^-$, where the nucleophilicities varies with the nature of L in the order: 2,6-lutidine > 2-picoline > pyridine > aniline > 4-cyanoaniline > Me_2S > $SbPh_3$ > $AsPh_3$ > PPh_3. This series reflects the decreasing σ-donor, and increasing π-acceptor strength of L, and hence the decreasing electron density at the metal centre.

Stereochemical change at the carbon

We have already mentioned the influence of the transition state geometry shown in Fig. 7.2 in defining the stereochemical consequences on the carbon centre. It is of paramount importance that the species whose stereochemistry is being measured is the initial product from the reaction, and not a complex that has undergone subsequent, secondary reactions which have destroyed any stereochemical integrity associated with the nucleophilic displacement process.

Certainly the most convincing evidence that the complex is acting as a nucleophile, via the two-centre transition state, is the demonstration that the carbon centre undergoes an inversion of configuration as a consequence of the oxidative-addition, as shown in Fig. 7.3.

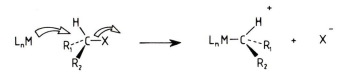

Fig. 7.3

The lack of crystallographic information confirming the absolute configurations of metal-alkyl complexes means that the determination of the stereochemistry at carbon is not a simple or direct experiment, and workers in this field have used a variety of ingenious methods to overcome this problem.

In the reaction of the iron complex and optically pure 2-bromobutane shown in Fig. 7.4, the initially formed alkyl complex is reacted with PPh_3 which induces the migration of the alkyl group onto one of the carbonyl ligands. As we shall see in Chapter 9 these migration reactions occur with retention of stereochemistry at the carbon centre.

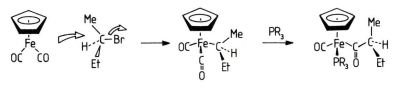

Fig. 7.4

Subsequent cleavage of the acyl group from the iron site is accomplished by Cl_2, and since this reagent cleaves the iron-carbonyl bond it does not affect the stereochemistry at the chiral carbon atom. In this way it was shown that the initial reaction of the iron nucleophile with the chiral alkyl bromide occurred with inversion of the configuration at the organic centre. In a similar manner the reactions of $[Pd(CO)(PPh_3)_2]$ with PhCHDBr have been shown to proceed with inversion of configuration at the carbon centre.

A direct means of determining the stereochemistry of an alkyl complex is to use an NMR spectroscopic method to distinguish between diastereoisomers. This has been accomplished for the nucleophiles $[Mo(\eta^5-C_5H_5)(CO)_3]^-$, $[Co(dmgH)_2py]^-$ and $[Fe(\eta^5-C_5H_5)(CO)_2]^-$ in their reaction with *erythro*-3,3-dimethylbutan-1-bromobenzenesulfonate-1,2-d_2, as shown in Fig. 7.5.

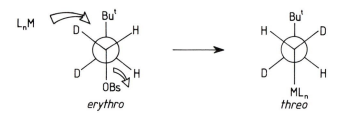

Fig. 7.5

The *threo-* and *erythro-*diastereoisomers can be distinguished by their characteristic vicinal coupling constants. In the reaction of the metal nucleophiles with the *erythro-*sulfonate, the product showed a single AB quartet (δ 1.38, J = 4.4 Hz) in the ^{1}H NMR spectrum. Independent preparation of the complex containing either the *threo-* or *erythro-*diastereoisomer confirmed that the product was the *threo-*isomer, J = 4.4 Hz, which was very different from the *erythro-*species, J = 13.1 Hz. Inversion at the carbon centre had been demonstrated.

So far, we have only addressed the stereochemistry of the carbon centre in its reactions with nucleophiles operating via the two-centre transition state. If the reaction operates through the three-centre transition state then one might expect that the reaction will occur with retention of configuration at the carbon centre. Although this remains an intriguing possibility, it has not so far been demonstrated.

7.2 Oxidative-addition by radical mechanisms

So far, we have seen a mechanism for the oxidative-addition reaction which is dominated by the nucleophilicity of the metal site. As a consequence, the reactivity towards alkyl halides with any particular nucleophile is in the order: RI > RBr > RCl; Me > Et > Pr^i > Bu^t, the same as is observed with the more conventional nucleophiles operating by an S$_N$2 mechanism. However, there are oxidative-addition reactions in which the order with respect to the alkyl group is the reverse of that shown above: PhCH$_2$ > Bu^t > Pr^i > Et > Me. Also, the oxidative-addition reactions of some complexes are not clean and significant amounts of products other than the alkyl complex are observed, as shown by the example in Fig. 7.6.

The rates of these reactions can be sensitive to traces of dioxygen, paramagnetic impurities or light. All these features point towards a free radical mechanism.

Detailed analysis of the reaction between [Co(CN)$_5$]$^{3-}$ and MeI demonstrated the reaction involved the rate-limiting halogen atom abstraction pathway shown in Fig. 7.7. The released methyl radical is scavenged rapidly by a second molecule of [Co(CN)$_5$]$^{3-}$.

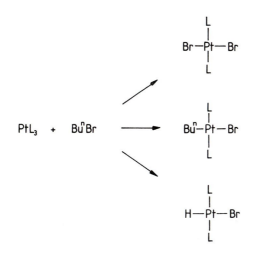

Fig. 7.6

$$[Co(CN)_5]^{3-} + MeI \longrightarrow [Co(CN)_5I]^{3-} + \overset{\cdot}{Me} \quad slow$$

$$[Co(CN)_5]^{3-} + \overset{\cdot}{Me} \longrightarrow [Co(CN)_5Me]^{3-}$$

Fig. 7.7

The analogous reaction with EtI produces significant amounts of $[Co(CN)_5H]^{3-}$ formed by hydrogen atom abstraction from the ethyl radical and the accompanying evolution of ethylene.

One of the most extensively studied systems that undergoes oxidative-addition by a free radical pathway is the reactions of $[Pt(PPh_3)_3]$ with RI shown in Fig. 7.8. This reaction gives essentially stoichiometric amounts of $[PtI(Me)(PPh_3)_2]$.

$$[Pt(PPh_3)_3] \longrightarrow [Pt(PPh_3)_2] + PPh_3$$

$$[Pt(PPh_3)_2] + RI \longrightarrow [PtI(PPh_3)_2] + \overset{\cdot}{R} \quad slow$$

$$[PtI(PPh_3)_2] + \overset{\cdot}{R} \longrightarrow [PtI(R)(PPh_3)_2]$$

Fig. 7.8

The reaction is rate-limited by iodine atom abstraction. The most important feature of this reaction is the detection of paramagnetic species which are formed after the rate-limiting step. In the presence of MeI, and the radical trap, Bu^tNO, strong signals attributable to the Pt(I) species and $Bu^tMeNO\cdot$ can be observed directly by EPR spectroscopy. These radicals are not derived from the individual reactants or from secondary reactions associated with the products.

Using 6-bromohex-1-ene as the substrate, further evidence for the radical nature of these reactions can be obtained. Using this alkyl bromide some of the product is the cyclopentylmethyl-platinum complex, formed from the rapid ($k = 1 \times 10^3$ s^{-1}) ring closing reaction of the hexene radical formed in the initial bromine atom abstraction step.

It is important to note that the analogous palladium complex, [Pd(PPh$_3$)$_3$], reacts with alkyl halides by means of the nucleophilic route; clearly the factors which discriminate between the two mechanisms are extremely subtle, and even in the reactions of [Pd(PEt$_3$)$_3$] with PhCHDBr at least part of the reaction goes by way of the radical mechanism.

Stereochemistry at the carbon atom

When the oxidative-addition reaction is accomplished by the nucleophilic pathway, inversion of the stereochemistry of the carbon centre is usually observed, as noted above. If the reaction operates via a pathway involving an alkyl radical it is clear that we would expect the free carbon radical to lose all 'memory' of its origins, and thus the reaction to be associated with loss of optical activity at the carbon centre.

This has been demonstrated in several systems, for instance in the reaction of *trans*-[IrCl(CO)(PR$_3$)$_2$] (PR$_3$ = PMe$_2$Ph, PMePh$_2$ or AsMe$_2$Ph) with MeCH(CO$_2$Et)Br and in the reaction of [Pd(PEt$_3$)$_3$] with PhCHDBr.

A system which demonstrates another general method of establishing the stereochemistry at a carbon centre using NMR spectroscopy is that employing the 2-substituted bromocyclohexanes, as shown in Fig. 7.9. A similar strategy can be adopted using 4-substituted bromocyclohexanes

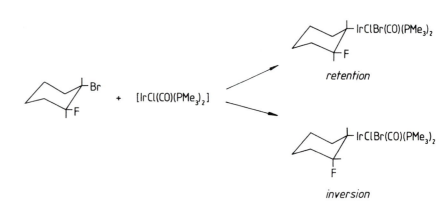

Fig. 7.9

This technique uses the principle that the bulky metal fragment prefers to enter a less congested equatorial site of the cyclohexyl ring, in so doing this defines the relative position of the 2- or 4-substituent. The stereochemistry of the product can be determined by comparison with the tert-butyl analogues, where again the bulky Bu^t residue prefers an equatorial site. In particular the magnitude of the coupling constants: J(ax-ax) = 8–14 Hz > J(eq–eq) or J(ax–eq) = 1.7 Hz are a good diagnostic tool.

7.3 Discrimination between nucleophilic and radical pathways

It is not a simple matter to know whether the radical or nucleophilic pathway is being followed in any particular oxidative-addition reaction. The relatively brief discussion of this chapter reveals that a metal complex can react with subtly different alkyl halides by different mechanisms, and changes in the ligands on the metal can also dramatically alter the reaction pathway. Clearly, determination of the stereochemistry of the reaction can unambiguously define the pathway, although this probe may not always be available to the experimenter. One simple test of the mechanism is to look at the reactions of alkyl tosylate, ROTs (TsO = *p*-toluenesufonate). This reagent is reluctant to react via the radical pathway, and hence a rapid reaction with MeI and MeOTs is indicative of a nucleophilic mechanism, whereas a rapid reaction with MeI but slow with MeOTs is more consistent with a free radical mechanism.

PART THREE: LIGAND-BASED REACTIONS

8 Activation of ligands

Having concentrated on reactions which centre on the metal, either in changing the oxidation state or changing a ligand, we will now briefly address the general reactivity of the ligand. This area of research defines many of the elementary reactions in catalysis involving metal sites, in particular, defining the action of metalloenzymes at the atomic level. For instance, studies on the cleavage of phosphates by metal ions give valuable insight into the hydrolysis of ATP in biochemistry. Also, understanding the factors controlling the attack of hydroxide at organic carbonyl compounds is of direct relevance to the action of the zinc-containing enzymes carboxypeptidase (which specifically cleaves peptides at the peptide carbonyl of the C-terminus) and carbonic anhydrase (which catalyses the equilibriation between CO_2 and H_2O). Similarly, understanding the factors that favour the protonation and reduction of coordinated dinitrogen, is important to the understanding of the action of the nitrogenases (which contain iron, together with molybdenum or vanadium and convert dinitrogen to ammonia by a sequence of electron and proton transfer reactions).

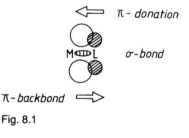

Fig. 8.1

8.1 Influence of the metal on the ligand

Simple consideration of the interaction between a metal and a ligand reveals that there are three types of bonding which we must consider, as shown in Fig. 8.1.

Besides the lowest energy interaction of σ-bonding from the ligand to the metal, the other interactions are π-bonding in nature either in the sense that electron density on the ligand is donated to the metal (π-donation) or the removal of any excess of electron density on the metal back to the ligand (π-backbonding). The relative magnitudes of these three interactions will define the final electron density on the ligand.

If the dominant characteristic of the metal is that it removes electron density from the ligand through both the σ-bonding and the π-donation characteristics, then the electron density on the ligand will be diminished compared with the situation where it is not bound to the metal. Hence the ligand will be more susceptible to attack by nucleophiles and any bonds within the ligand may be more prone to cleavage. This type of activation of ligands is most associated with 'hard' metal sites, that is a relatively high formal oxidation state and often a first row transition metal, making complexes that are best described as being electron-poor. The metal that dominates this area of chemistry mechanistically is cobalt(III).

In contrast, if the dominant characteristic of the metal is to push electron density onto the ligand via the π-backbonding then the increased electron density on the coordinated ligand will result in an increased susceptibility to attack by electrophiles and an increased nucleophilicity of the ligand. This type of bonding is most associated with 'soft' metal sites, that is a relatively low formal oxidation state and often a second or third row transition metal. These complexes are described as being electron-rich.

Of course the electron density distribution in metal complexes is not as clear cut as implied by the above discussion, and the relative contributions from each type of bond, results in a graduation of reactivities associated with the ligand. For instance, the d^6 'Co$(NH_3)_5^{3+}$' activates coordinated esters towards hydroxide attack, whereas 'Ru$(NH_3)_5^{2+}$' (which is also a d^6 metal) is much less efficient at activating esters in this fashion, but will readily bind dinitrogen: a ligand which binds extensively by π-backbonding. However, the ruthenium site does not release sufficient electron density onto dinitrogen to activate it towards attack by protons. In order to activate dinitrogen towards protic attack we must move to a much more electron-releasing d^6 site, the 'Mo$(Ph_2PCH_2CH_2PPh_2)_2$' core.

8.2 Ligand activation by electron-poor sites

In this section we will discuss briefly the two dominant modes of activation of ligands at electron-poor sites: heterolytic cleavage of bonds within the ligand as exemplified by the studies on substituted phosphates, and attack by nucleophiles at the ligand as exemplified by the attack of hydroxide at the esters and amides of coordinated amino acids.

Cleavage of phosphates

The hydrolysis of substituted phosphates occurs by means of a dissociative pathway, as shown in Fig. 8.2.

Binding of the phosphate to a simple labile metal site such as Mn^{2+}, Fe^{n+} (n = 2 or 3), Co^{2+}, Ni^{2+} or Cu^{2+} (all 'hard' sites) results in the rapid cleavage of the phosphate, as shown in Fig. 8.2. This increased rate of dissociation upon coordination of the metal is attributable to the metal withdrawing the electron density from the phosphorus–oxygen bond undergoing cleavage. The influence of the metal is entirely one involving polarization of the electron density within the ligand.

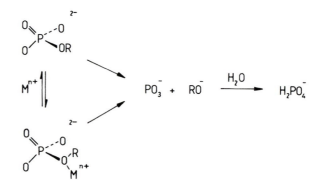

Fig. 8.2

Nucleophilic attack at ligands

The hydrolysis of peptides, esters, amides, etc., as shown in Fig. 8.3, is catalysed by labile 'hard' metal ions such as Co^{2+} or Cu^{2+}, and has been known since the early 1950s.

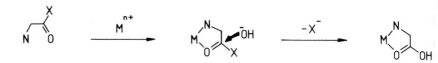

Fig. 8.3

The generally accepted mechanism for this catalysis as shown in Fig. 8.3 was described in these early studies and remains to this day, essentially unchanged. The formation of the five-membered chelate ring is favoured by the large number of labile sites available on these metal centres. The binding of the peptide to the 'hard' metal ion renders the carbonyl carbon atom more susceptible to hydroxide attack since the metal pulls electron density away from this carbon. This type of activation of the carbonyl functional group is quite general and similar to the electron-withdrawing effect we saw in the hydrolysis of phosphates.

The uncertainty concerning the binding of the ligand in the studies on these substitutionally labile metal sites can be circumvented by using substitutionally robust complexes. By using simple cobalt(III) complexes, such as that shown in Fig. 8.4, the hydrolysis of amino acid derivatives has been defined.

In the reactions shown, the linear quadridentate trien ligand $(H_2NCH_2CH_2NHCH_2CH_2NHCH_2CH_2NH_2)$ acts as a spectator to the reactions occurring at the other two sites, and does not even change its geometrical orientation during the course of the reaction. By using this robust system it is possible to:

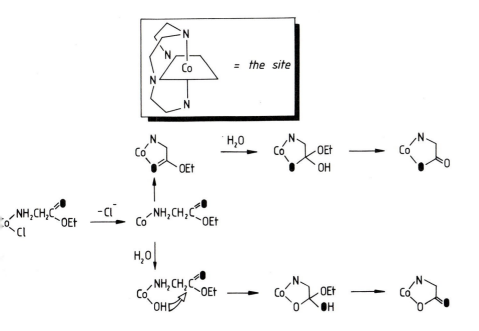

Fig. 8.4

(1) introduce isotopic labels at specific positions in the complex and hence identify the details of the process;

(2) establish the geometries of the reactants and any detectable intermediates.

The robust chloro-complex shown in Fig. 8.4 can be activated rapidly by removal of the chloro-group using either Hg^{2+} or HO^- (by a D_{cb} mechanism). The site vacated by the chloro-group can then be attacked by either the carbonyl oxygen atom of the ester to form the ring-closed species (top pathway) or by hydroxide (bottom pathway). In the D_{cb} reaction, both pathways are observed. Subsequent hydrolysis of the carbonyl group occurs by:

(1) the intermolecular attack of hydroxide at the carbonyl carbon of the ring-closed species, or

(2) by intramolecular attack of the coordinated hydroxide on the ester, as shown in the bottom line of Fig. 8.4.

Clearly, in a system based on a labile metal site these two pathways would be indistinguishable. However in the cobalt(III) study it is possible to start with ^{18}O-labelled glycine ester with the isotopic label specifically in the carbonyl oxygen atom. Now the products of the two pathways can be distinguished. For the intermolecular pathway (involving the ring-closed ester) the product will contain the isotopic label within the chelate ring, whereas in the intramolecular pathway the isotopic label will remain uncoordinated to the cobalt.

A similar distinction is possible using ^{18}O-labelled water.

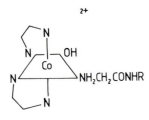

Fig. 8.5

The hydrolysis of carbonyl compounds is quite general and much work has also been done on the analogous amide systems, such as that shown in Fig. 8.5. The hydrolysis of the glycine amides occurs about 10^5 times more slowly than the reactions of glycine esters and so it has been possible to isolate the ring-closed species analogous to that shown in Fig. 8.4.

In the reactions of the glycine amides, the coordinated hydroxide reacts about 10^2 times faster than external attack of free hydroxide on the chelated amide. Interestingly, coordinated water reacts even more rapidly than coordinated hydroxide.

At first sight it is a little surprising that the hydroxide coordinated to cobalt(III) is such a good nucleophile in these systems; the cobalt(III) centre is a 'hard' metal ion and thus a coordinated hydroxide ion would be expected to be significantly less nucleophilic than free hydroxide, since the metal will drain the hydroxide of electron density. Certainly a hydroxide coordinated to cobalt(III) is about 10^8 times less basic than free hydroxide ion. Despite this, the intramolecular pathway, is about 10^9 times faster than the intermolecular route for the reactions of the amides. This increased activity of the intramolecular pathway is due to the coordinated hydroxide being in the correct position, at the correct time to accomplish the hydrolysis reaction. In the intermolecular pathway the hydrolysis of the carbonyl residue has to wait for the arrival of the hydroxide ion from the bulk solvent. An important feature of metal-mediated reactions is the availability of several coordination sites on the same metal to bring reagents close together and hold them in position until the reaction is ready to proceed.

Thus we see that the role of the cobalt centre in these hydrolysis reactions is not only to polarize the carbonyl group and make it more susceptible to nucleophilic attack, but also to coordinate the nucleophile so that it is in the correct position to accomplish the hydrolysis.

A further reaction in which a coordinated hydroxide is involved is shown in Fig. 8.6.

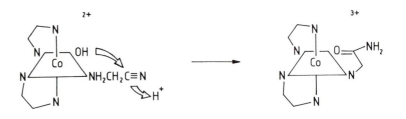

Fig. 8.6

Again we observe a dramatic increase in the rate of hydrolysis of coordinated nitriles of about 10^{11} compared to the free species.

8.3 Ligand activation by electron-rich sites

Binding of small molecules to electron rich metal sites can result in an increased basicity of that molecule, and hence its ready protonation under conditions where the free molecule is only weakly protonated or not protonated at all. In the same way that activation of ligands at electron-poor sites is dominated by cobalt(III) complexes, the activation of ligands at electron-rich sites is dominated by reactions at the 'M(diphosphine)$_2$' core (M = Mo, W, Fe, etc.), at least from a mechanistic point of view. As shown in Fig. 8.7, molecules as diverse as isonitriles, unsaturated hydrocarbons and dinitrogen are all readily protonated when coordinated to this site.

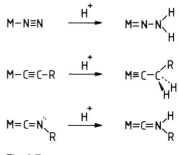

Fig. 8.7

Of these examples, the system most well examined mechanistically is the diprotonation of coordinated dinitrogen.

Protonation of coordinated dinitrogen

The reaction of the complexes *trans*-[M(N$_2$)$_2$(Ph$_2$PCH$_2$CH$_2$PPh$_2$)$_2$] (M = Mo or W) with HX (X = Cl, Br or I) in tetrahydrofuran results in the diprotonation of one of the coordinated dinitrogen residues to form the hydrazido(2-)-complex, as shown in Fig. 8.7. The exact mechanistic pathway chosen depends upon the nature of both the acid and the metal. The general mechanism is shown in Fig. 8.8.

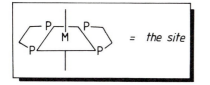

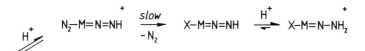

Fig. 8.8

The simplest pathway involves monoprotonation of one of the dinitrogen ligands. This has the effect of withdrawing electron density from the metal centre and hence weakening the bonding of the *trans* dinitrogen ligand, which subsequently dissociates from the metal centre in the rate-limiting step of the reaction. Attack of halide ion at the vacant site generates the diazenido-complex. The coordinated halide is significantly more electron-releasing than the dinitrogen molecule that has just departed and hence increases the basicity of the diazenido-ligand sufficiently to bind the second proton rapidly and form the product. An important feature of these reactions is that the acts of protonation of the dinitrogen residues are extremely rapid, being diffusion-controlled or close to the diffusion-controlled limit.

At relatively high concentrations of acid, or with the stronger acids (remember that in the aprotic solvents the hydrohalic acids are not 'levelled', that is HX is not completely dissociated into H^+ and X^-), increasing contributions from a pathway involving diprotonation of the dinitrogen ligand prior to the rate-limiting dissociation is observed. In addition, the contribution from the diprotonation pathway is always more prominent with the tungsten complex than in the analogous reactions with the molybdenum species. This is a consequence of the greater electron-releasing nature of the heavier elements in the group.

A problem which recurs in studying the protonation reactions of electron-rich sites is the formation of hydrido-complexes formed by proton attack at the metal. This occurs in the reactions based on the '$M(Ph_2PCH_2CH_2PPh_2)_2$' core in the reactions with HCl, which ultimately results in the deactivation and dissociation of dinitrogen. However, protonation at the metal need not always lead to deactivation of the dinitrogen and just a subtle change in the reaction site can change the mechanism. As shown in Fig. 8.9, the reaction of *trans*-$[Mo(N_2)_2(Et_2PCH_2CH_2PEt_2)_2]$ with HCl results in the formation of the hydrazido(2-)-complex but now the initial intermediate detected is the hydrido-species.

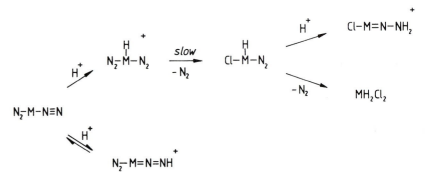

Fig. 8.9

The ethyl-substituted phosphine is more electron-releasing than the corresponding phenyl-derivative. Protons will still bind to the dinitrogen ligand in *trans*-$[Mo(N_2)_2(Et_2PCH_2CH_2PEt_2)_2]$ at the diffusion-controlled rate, but once bound to the dinitrogen they do not labilize the *trans* dinitrogen ligand sufficiently because of the strong electron-releasing capability of the ethyl-substituted phosphine. Consequently, although protons are rapidly binding and dissociating from the dinitrogen ligand the coordination sphere of the complex remains intact. Now the slower protonation of the metal can occur. Protonation at the metal is much slower than the diffusion-controlled limit, in part because of the diffuse nature of the electron density at the metal (there is no stereochemical lone pair of electrons as there is on a terminal nitrogen atom). Protonation at the metal can only occur under conditions where protonation at the dinitrogen is non-productive, as in this case. Protonation of the metal has a strong labilizing effect on the proximal dinitrogen ligands, and rate-limiting dissociation can now occur. As we saw in the previous example, binding of the more electron-releasing chloro-group to the metal renders the remaining dinitrogen residue sufficiently basic that, at high concentrations of HCl, it is able to pick up protons and proceed through to the hydrazido(2-)-product, after dissociating the hydrido-ligand as a proton. At low concentrations of acid the second dinitrogen dissociates before it can be protonated.

Studies on the other small molecules bound to the 'M(diphosphine)$_2$' core shown in Fig. 8.7, indicates that similar factors are involved in the protonation of these residues to those described for dinitrogen. The only difference appears to be a more prevalent intramolecular pathway, involving migration of hydrido-groups onto the ligand. Despite many attempts to observe migrations in the reactions of the type shown in Fig. 8.8 and Fig. 8.9, it appears that the reaction of dinitrogen, at least in the initial protonation steps, is dominated by proton addition and dissociation steps.

Dialkylation of dinitrogen ligands

Despite the superficial similarity between protonation and alkylation of ligands bound to electron-rich sites, studies on the dinitrogen complexes makes it clear that the mechanism of one reaction cannot be used to imply the mechanism of the other. As we have seen in Chapter 7, alkyl halides can avoid being attacked by nucleophiles, and this is what happens in the reactions with complexes of the type *trans*-[M(N$_2$)$_2$(diphosphine)$_2$]. The mechanism for the initial alkylation of the dinitrogen ligand involves rate-limiting dissociation of a dinitrogen ligand, as shown in Fig. 8.10.

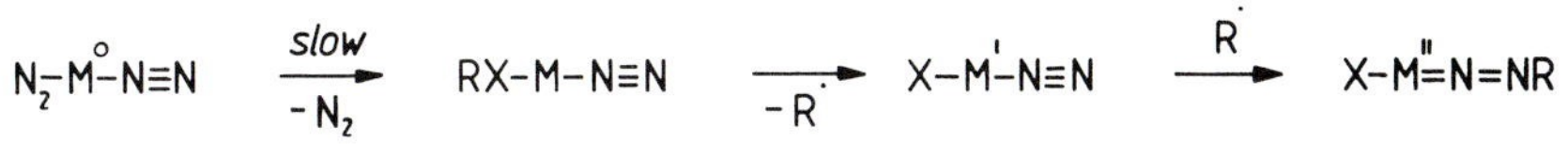

$$N_2\text{--}\overset{o}{M}\text{--}N{\equiv}N \xrightarrow[-N_2]{slow} RX\text{--}M\text{--}N{\equiv}N \xrightarrow[-R^{\cdot}]{} X\text{--}\overset{\iota}{M}\text{--}N{\equiv}N \xrightarrow{R^{\cdot}} X\text{--}\overset{\shortmid\shortmid}{M}{=}N{=}NR$$

Fig. 8.10

Once the dinitrogen ligand has departed the alkyl halide, RX, can bind to the vacant site via the halide and subsequently undergoes homolytic fission to give the alkyl radical, which then rapidly attacks the dinitrogen ligand to generate the monoalkylated species. This mechanism is the atom abstraction process we have met before in Chapters 6 and 7.

In contrast to the non-chain, radical pathway for the addition of the first alkyl group to a dinitrogen ligand, the second alkylation step involves a simple nucleophilic attack of the alkyldiazenido-residue on an alkyl halide, as shown in Fig. 8.11.

Fig. 8.11

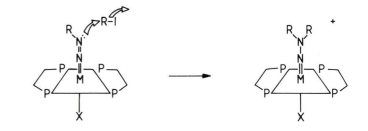

This mechanism we have also met before in oxidative-addition reactions. This recurring comparison with the mechanisms observed for the oxidative-addition reactions and the dialkylation of these dinitrogen complexes is no coincidence; the alkylation of dinitrogen complexes can be considered as an oxidative-addition process since the metal increases its formal oxidation state by two units for each addition of an alkyl halide molecule. The only difference between the reactions discussed here and those discussed in Chapter 7 is that in the present case, the alkyl group adds to a ligand rather than to the metal.

9 The insertion reaction

In this chapter we will discuss a general class of reaction whose title is mechanistically misleading, that is the insertion reaction. The overall effect of the reaction is shown in Fig. 9.1, but this bears no relationship to the mechanism. This reaction is accomplished by a variety of small molecules, most notably sulfur dioxide, alkynes, alkenes, carbon monoxide, and nitric oxide. These molecules can be inserted into metal–carbon bonds and sometimes into metal–hydrogen bonds.

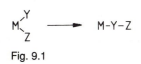

Fig. 9.1

In most cases it has been demonstrated that the mechanistic description of the reaction is that of an intramolecular migration process. However, in at least one case, the reaction of CO with $[Fe(\eta^5\text{-}C_5H_5)(CH_3)(CO)_2]$, a direct insertion into the Fe–CH$_3$ bond has been observed; but we will concentrate on the migration reactions.

In what follows there is no attempt to discuss the mechanistic studies on all the substrates listed above, rather a discussion of the general principles of the mechanism as they relate to the last three molecules: alkenes, carbon monoxide and nitric oxide.

9.1 Insertion of carbon monoxide

The first mechanistic study of the reaction of an insertion involving carbon monoxide was that shown in Fig. 9.2.

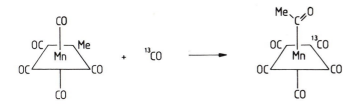

Fig. 9.2

By use of the labelled ^{13}CO it can be shown that the attacking molecule of carbon monoxide is not the group that undergoes the insertion reaction, rather it is one of the carbon monoxide molecules already coordinated to the manganese. Also, other groups such as PPh$_3$ can facilitate the migration reaction. Irrespective of the nature of the molecule being added to the metal site, the immediate product of the reaction is the isomer with the added molecule, ^{13}CO or PPh$_3$, cis to the acyl-residue. This is what

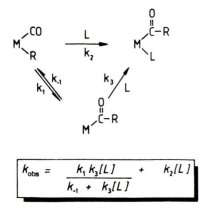

$$k_{obs} = \frac{k_1 k_3 [L]}{k_{-1} + k_3 [L]} + k_2 [L]$$

Fig. 9.3

would be expected if the migrating group (methyl residue) and the migration terminus (carbon monoxide ligand) must be adjacent to one another, as shown in the generalized mechanism in Fig. 9.3.

The pathways shown in Fig. 9.3 illustrate the two limiting mechanisms for the migration reaction: the pathway in which the migration step occurs before the addition of the group L, and the bimolecular pathway in which the group L enters the coordination sphere as the migration occurs. The analogy between these two routes and the associative and dissociative substitution pathways we met in Part 2 of this book is obvious. Of course, under the experimental conditions used to investigate the mechanism of the reaction, the observed rate law is often much simpler than the generalized form shown in Fig. 9.3.

In the example shown in Fig. 9.4 the rate of the reaction exhibits a simple first order dependence on the concentration of the complex and is independent of the concentration and nature of the nucleophile. This is consistent with the reaction operating by the stepwise pathway exclusively with $k_3[L] \gg k_{-1}$.

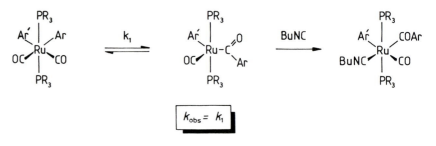

$$k_{obs} = k_1$$

Fig. 9.4

In another system shown in Fig. 9.5, when X = I, Br or Cl, the rate of the reaction exhibits a simple first order dependence on the concentration of both the complex and PPh_3. This is consistent with the stepwise mechanism shown provided that $k_{-1} \gg k_3[PPh_3]$.

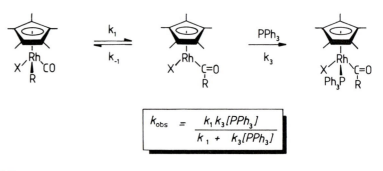

$$k_{obs} = \frac{k_1 k_3 [PPh_3]}{k_1 + k_3 [PPh_3]}$$

Fig. 9.5

However, when X = $MeCO_2$ or CF_3CO_2, the kinetics obey the full form of the rate law, as shown in Fig. 9.5. It is proposed that the carboxylate groups act as bidentate ligands in the coordinatively unsaturated intermediate and suppress the k_{-1} pathway, ensuring that at high concentrations of PPh_3, $k_3[PPh_3] \geqslant k_{-1}$.

Solvent effects

Some migration reactions are strongly influenced by the solvent. Thus in the reaction of $[MnMe(CO)_5]$ with $C_6H_{11}NH_2$ the rate of the reaction varies dramatically with the dielectric constant and coordinating power of the solvent such that the rate of the reaction in N,N'-dimethylformamide is 10^4 times faster than in mesitylene. This influence of the solvent has ramifications in terms of the intimate mechanism for the migration reaction. In particular the existence of a true unassisted migration reaction seems unlikely. It is more likely that if the group L cannot assist the migration, then the large excess of solvent molecules will do so. This has been demonstrated in the system shown in Fig. 9.6.

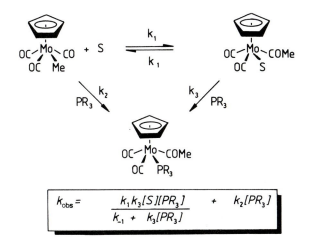

$$k_{obs} = \frac{k_1 k_3 [S][PR_3]}{k_{-1} + k_3[PR_3]} + k_2[PR_3]$$

Fig. 9.6

In this system the migration reaction is observed to be promoted not only by the tertiary phosphine in the direct pathway but also by the solvent (S), tetrahydrofuran, which is subsequently replaced by the phosphine. When the solvent is changed to 2,5-dimethyltetrahydrofuran there is little change in the dielectric constant of the reaction mixture, but the methyl substituents are too bulky to permit strong binding of this molecule to the metal and thus $k_3[PR_3] \ll k_{-1}$ and the kinetics maintain the first order dependence on the concentration of phosphine in the concentration ranges responsive to study. The role of the attacking solvent molecule is not just to stabilize the coordinatively unsaturated intermediate, but also to catalyse the formation of the unsaturated intermediate. This has been demonstrated in the reaction shown in Fig. 9.7.

$$(CO)_5Mn-Bs \; + \; \underset{\longleftarrow}{\overset{S}{\longrightarrow}} \; (CO)_4SMn-C\overset{\nearrow O}{\underset{\searrow Bs}{}} \; \underset{\longleftarrow}{\overset{-S}{\longrightarrow}} \; (CO)_4 \, Mn-C\overset{\nearrow O}{\underset{\searrow Bs}{}}$$

$$Bs = CH_2 \text{—} \hspace{-0.5em}\bigcirc\hspace{-0.5em}\text{—} OMe$$

Fig. 9.7

Detailed kinetic analysis of this reaction demonstrates that the solvent molecule dissociates from the metal site having accomplished its role of 'pushing' the migration step.

Movement of the alkyl group

One of the most important questions to be addressed in insertion reactions is, which group moves in the migration? There are three possibilities, as shown in Fig. 9.8, the group R moves on to a static CO (top), CO moves towards a static R (bottom) or both move towards one another (middle).

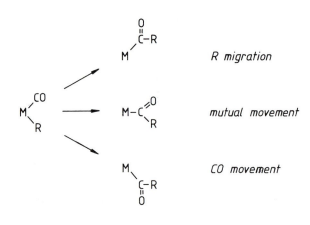

Fig. 9.8

The product distribution from the reaction of the isotopically labelled *cis*-[MnMe(^{13}CO)(CO)$_4$] with CO, shown in Fig. 9.9, demonstrates that the intimate mechanism is one in which the methyl group migrates onto an adjacent CO group. The product distribution is inconsistent with a mechanism involving the movement of the CO ligand towards the Me residue, and if there was a co-operative movement of the two ligands towards one another then an increased ratio of the *cis*-[Mn(COMe)(^{13}CO)(CO)$_4$] to the *trans* isomer would be observed.

That it is the alkyl group that moves, is consistent with the retention of stereochemistry at the carbon centre during the insertion reaction. This has been shown by demonstrating that the optical activity of the equilibrium mixture illustrated in Fig. 9.10 does not change with time.

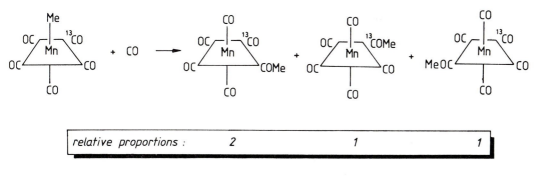

Fig. 9.9

Fig. 9.10

In addition, the observation that many Lewis acids increase the rate of insertion by up to a factor of 10^8 is readily rationalized in terms of the migration of the alkyl group. If the Lewis acid binds to the oxygen atom of the carbonyl group, as shown in Fig. 9.11, then it will pull electron density towards itself, thus rendering the carbonyl carbon atom more positively charged and accelerating the movement of the migrating electron-rich alkyl residue.

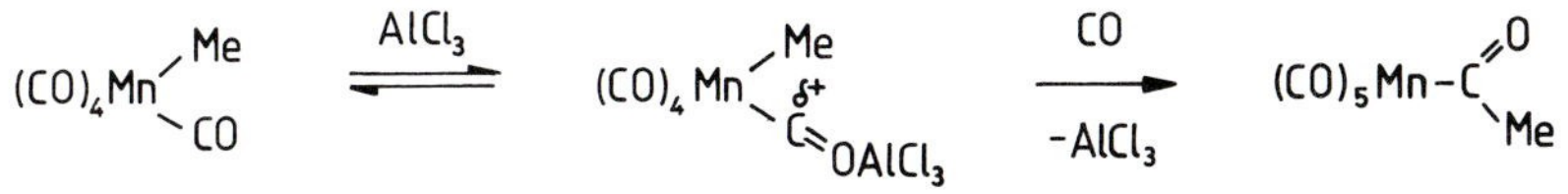

Fig. 9.11

Stereochemistry at the metal

The stereochemical consequences of the insertion reaction on the metal centre has been little studied. However, in one study, shown in Fig. 9.12, the stereochemistry of the reverse reaction (the retro-insertion) has been established. We can use the principle of microreversibility to predict that the migration reaction at the same site occurs with inversion of configuration of the iron. The principle of microscopic reversibility is a useful tool in mechanistic elucidation. In its simplest and broadest terms ,the principle requires that the mechanism in the forward reaction is the same as in the reverse reaction. That is, the position of each atom at any time during the reaction, in either direction, occupies the same relative positions.

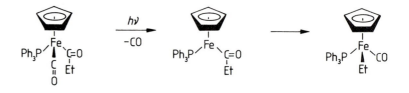

Fig. 9.12

In order to accommodate the inversion of configuration at the iron site in the retro-insertion reaction, it is clear that the coordinatively unsaturated intermediate must maintain its configuration or, if a solvent molecule occupies the site, the alkyl group migrates into its position.

Influence of the migrating group

It is difficult to quantify the influence of the migrating group on the rate of the reaction, free from the complications of the other parameters. The most extensive study is on the migration reactions of niobocene hydrides or alkyls shown in Fig. 9.13.

Fig. 9.13

$$Cp_2Nb{=}C\underset{R}{\overset{H}{{<}}}\underset{O{-}ZrHCp_2^*}{} \rightleftharpoons Cp_2Nb{-}C\overset{H}{\underset{O{-}ZrHCp_2^*}{{<}}}R \xrightarrow{L} Cp_2Nb{-}C\overset{H}{\underset{L}{{<}}}\underset{O{-}ZrHCp_2^*}{}R$$

The rates of these reactions were measured directly by an NMR spectroscopic technique, spin-saturation-transfer experiments. The order in which the groups migrate is H $\gg$ Me > CH$_2$Ph. The much greater propensity of the hydride group to migrate reflects the greater flexibility of the non-directional s orbital employed in the three-centre, two electron bond that represents the transition state of the migration reaction. With alkyl-groups migrating, the directional nature of the p orbitals imposes constraints on the transition state.

Of course, the relative migrating ability of the alkyl groups varies with the nature of the metal site. Thus in the migration of [Rh(η^5-C$_5$H$_5$)(I)(R)(CO)] the R = Me residue migrates four times slower than R = Ph, whereas for [MnR(CO)$_5$] the R = Me residue migrates eight times faster than R = Ph.

Redox-induced migratory insertion

Using transient electrochemical techniques, the insertion reaction of the 17 electron species, [Fe(η^5-C$_5$H$_5$)Me(CO)(PPh$_3$)]$^+$ with pyridine has been studied. The mechanism is that shown in Fig. 9.14(A) in which the pyridine binds to the metal centre to catalyse the migration of the methyl group. Thus, the reaction involves the intermediacy of a nineteen electron species.

Fig. 9.14 (A)

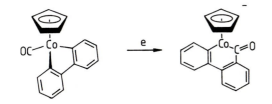

Fig. 9.14 (B)

In Chapter 5 we saw that the substitution reactions of 17 electron species invariably proceeds via a 19 electron intermediate. The same mechanistic pattern is observed in the electrochemical reduction of the 18 electron cobalt complex shown in Fig. 9.14(B). Reduction in this case results in the rapid insertion reaction for the 19 electron species.

9.2 Insertion of nitric oxide

Despite the superficial similarity between carbon monoxide and nitric oxide, insertion reactions of the latter molecule are rare. Detailed mechanistic study of the insertion of nitric oxide demonstrates marked differences between the mechanism for nitric oxide and carbon monoxide.

The mechanism of the insertion of NO into the methyl group of $[Co(\eta^5\text{-}C_5H_5)Me(NO)]$ is shown in Fig. 9.15.

In this mechanism it is the linear nitric oxide ligand which undergoes the migration reaction, in the pathway involving subsequent attack of L = PPh_3. Most importantly, the use of the more nucleophilic L = PEt_3 results in direct attack of this phosphine at the metal. Rather than pushing the methyl group to migrate to the nitric oxide ligand, the complex relieves the formal 20 electron count at the metal by the linear nitric oxide becoming bent. The bent form of nitric oxide is incapable of undergoing the insertion reaction at an appreciable rate and must await its reconversion to $[Co(\eta^5\text{-}C_5H_5)Me(NO)]$ before migration can occur.

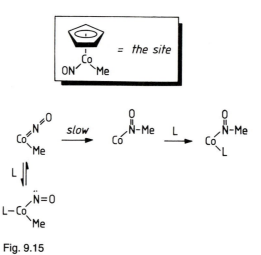

Fig. 9.15

9.3   Insertion of alkenes

Despite the obvious importance of the insertion of alkenes to many catalytic processes, relatively little has been defined in terms of the intimate details of the mechanism. However, many of the general features established for the analogous reactions of carbon monoxide are operative. Thus in the study of the reversible insertion of *trans*-[RhH(η^2-C_2H_4)(PPr$_3^i$)$_2$] by magnetization-transfer techniques, the essential prerequisite that the *trans* complex rearranges to the *cis* isomer is observed for the reaction to proceed by an intramolecular pathway.

The most extensive study of the insertion reaction of alkenes, free from the complications of subsequent processes, is in the general system shown in Fig. 9.16.

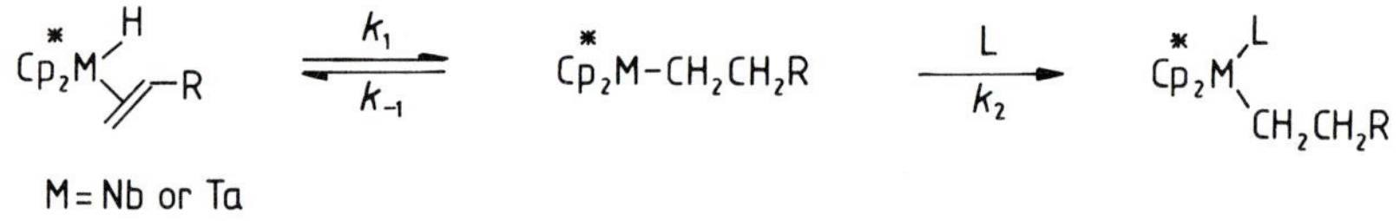

M = Nb or Ta

Fig. 9.16

In this system attempts to determine the value of k_1 by a kinetic method (trapping the inserted product by a nucleophile L) were unsuccessful since with L = CO or MeNC, $k_2[L] \ll k_{-1}$ under all experimental conditions, resulting in simple first order dependence on the concentration of L. However, it is possible to determine the value of k_1 directly by magnetization-transfer and coalescence techniques. This leads to a proposed four-centre transition state for the insertion, where the hydride group migrates as H$^-$. By coordinating a $(C_\alpha)H_2(C_\beta)HPh$ styrene and varying the nature of the *para*-substituents on the phenyl ring, it has been possible to delineate the electronic effects of the migration reaction as follows.

(1) Electron-withdrawing groups at C_α or C_β stabilize the ground state of the reactant by increasing the binding efficiency of the alkene to the niobium.

(2) Electron-releasing groups at C_β accelerate the rate of migration by stabilizing the incipient positive charge at this carbon atom in the transition state.

(3) Electron-withdrawing groups at C_α also accelerate the rate of insertion, probably by increasing the bonding between niobium and C_α in the transition state.

These mechanistic conclusions are greatly aided by the results of an X-ray crystal structure on $[Nb(\eta^5\text{-}C_5Me_5)_2H(CH_2CHPh)]$ which shows that the phenyl group is orientated so that the effect of a substituent on this ring is not transmitted to the carbon–carbon double bond in the ground state. However upon approaching the transition state the phenyl ring must twist back into conjugation with the alkene bond in order that the substituents influence the reaction rate.

Additional reading

General

Burgess, J. (1988). *Metal Ions in Solution*, Wiley, New York.

Espenson, J. H. (1981). *Chemical Kinetics and Reaction Mechanisms*, McGraw Hill, New York.

Kochi, J. K. (1978). *Organometallic Mechanisms and Catalysis*, Academic Press, New York.

Wilkins, R.G. (1991). *Kinetics and Mechanism of Reactions of Transition Metal Complexes*, VCH, Weinheim.

Moore, J. W. and Pearson, R. G. (1981). *Kinetics and Mechanism*, Wiley, New York.

Atwood, J.D. (1985). *Inorganic and Organometallic Reaction Mechanisms*, Brooks/Cole, Monterey.

Substitution at square-planar sites

Cross, R. J. (1989). *Advances in Inorganic Chemistry*, **34**, 219–292.

Substitution mechanisms

Tobe, M. L. (1987). *Comprehensive Coordination Chemistry* (ed. G. Wilkinson, R. D. Gillard, and J. A. McCleverty), **1**, Chapter 7.1, Pergamon, Oxford.

Howell, J. A. S. and Burkinshaw, P. M. (1983). *Chemical Reviews*, **83**, 557–599.

Catalysed substitution

Lawrance, G. A. (1989). *Advances in Inorganic Chemistry*, **34**, 145–194.

Tobe, M. L. (1983). *Advances in Inorganic and Bioinorganic Mechanisms*, **2**, 1–94.

Electron transfer

Cannon, R. D. (1980). *Electron Transfer Reactions*, Butterworths, London.

Haim, A. (1983). *Progress in Inorganic Chemistry*, **30**, 273–303.

Marcus, R. A. (1964). *Annual Reviews in Physical Chemistry*, **15**, 155–196.
Meyer, T. J. (1983). *Progress in Inorganic Chemistry*, **30**, 389–412.
Meyer, T. J. and Taube, H. (1987). *Comprehensive Coordination Chemistry* (ed. G. Wilkinson, R. D. Gillard, and J. A. McCleverty), **1**, Chapter 7.2, Pergamon, Oxford.
Taube, H. and Gould, E. S. (1969). *Accounts of Chemical Research*, **2**, 1225.

Oxidative-addition

Henderson, S. and Henderson, R. A. (1987). *Advances in Physical Organic Chemistry*, **23**, 1–62.
Lappert, M. F. and Lednor, P. W. (1976). *Advances in Organometallic Chemistry*, **14**, 345–399.

Reactions at ligands

Hay, R. W. (1987). *Comprehensive Coordination Chemistry* (ed. G. Wilkinson, R. D. Gillard, and J. A. McCleverty), **6**, *Chapter 61.4*, *Pergamon, Oxford.*

Insertion reactions

Anderson, G. K. and Cross, R. J. (1984). *Accounts of Chemical Research*, **17**, 67–74.
Wojcicki, A. (1974). *Advances in Organometallic Chemistry*, **12**, 31–81.